32nd Annual Meeting of the North American Membrane Society (NAMS 2023)

A New Era of Sustainable Membrane Innovation

Tuscaloosa, Alabama, USA
13 – 17 May 2023

Editor:

Steven T. Weinman

ISBN: 978-1-7138-8498-9

Printed from e-media with permission by:

Curran Associates, Inc.
57 Morehouse Lane
Red Hook, NY 12571

Some format issues inherent in the e-media version may also appear in this print version.

North American Membrane Society (NAMS)
University of Arkansas
Ralph E. Martin Department of Chemical Engineering
3202 Bell Engineering Center
Fayetteville, AR 72701-1201
USA

Phone: +1 479 575-7926
Fax: +1 479 575-7926

names@uark.edu

Additional copies of this publication are available from:

Curran Associates, Inc.
57 Morehouse Lane
Red Hook, NY 12571 USA
Phone: 845-758-0400
Fax: 845-758-2633
Email: curran@proceedings.com
Web: www.proceedings.com

NORTH AMERICAN MEMBRANE SOCIETY

32nd ANNUAL MEETING

2023

MAY 13TH – 17TH

EMBASSY SUITES, TUSCALOOSA, AL

A NEW ERA OF SUSTAINABLE MEMBRANE INNOVATION

www.membranes.org/nams-2023/

Program Book

Conference chairs
Milad R. Esfahani
Steven T. Weinman
Jason E. Bara
The University of Alabama

Edited by
Steven T. Weinman

Table of Contents

Welcome to NAMS 2023

On behalf of the NAMS Board of Directors, it is our pleasure to welcome you to NAMS 2023 in Tuscaloosa, Alabama!

We are excited to gather together once again to continue our shared pursuit of membrane science and technology. This year's meeting promises to be an exciting and informative event, and we would like to thank our chairs, Milad Esfahani, Steven Weinman, and Jason Bara for their outstanding efforts in organizing this meeting and creating a program that is both engaging and timely.

The theme for NAMS 2023 is "A New Era of Sustainable Membrane Innovation". We have an outstanding lineup of plenary speakers (James F. Reilly, Lidietta Giorno, Nick Irvin), as well as a diverse array of technical sessions covering topics such as membrane materials and fabrication, water and energy applications, biotechnology and medical applications, and modeling and simulation.

As always, NAMS 2023 provides a platform for scientific exchange, collaboration, and networking. We encourage all attendees to take advantage of the many opportunities to engage with colleagues and experts in the field, including the poster session, exhibitor displays, and various social events.

We are also proud to continue our commitment to supporting the next generation of membrane scientists and engineers. The workshops, industry-student mentorship program, and Elias Klein Founders' travel supplements are just a few of the initiatives that NAMS offers to help support and mentor students and early-career professionals in the field.

Finally, we would like to take this opportunity to recognize the achievements of our distinguished members. At this year's meeting, we will induct two new NAMS Fellows for their contributions to the field and service to the society: Isabel Escobar and Glenn Lipscomb. We congratulate them on this well-deserved honor.

Thank you for joining us in Tuscaloosa, Alabama for NAMS 2023. We look forward to a successful and productive meeting, and to the continued growth and success of the membrane science and technology community.

Ryan Lively, NAMS President and Marie-Eve Langevin, NAMS Vice President

Message from the 2023 NAMS Conference Chairs

It is our pleasure to welcome you to Tuscaloosa for the 32nd Annual Meeting of the North American Membrane Society! It has been our honor to serve as co-chairs for this year's NAMS meeting.

We will use Whova as a platform for program information, message distribution, and networking, so make sure you download the app on your phone and bookmark it on your computer. We will have ~125 posters and 170 talks on a wide range of membrane-related topics. These contributions are of the highest quality and we are sure to have an engaging conference.

We thank our dedicated oral and poster session chairs. Also special thanks to the NAMS 2023 Conference supporter (Manish Kumar, Mary Laura Lind, John Pellegrino, Steve Ritchie, and Heath Turner) and NAMS board members and leadership (especially Jamie Hestekin, Lucy Camacho, and Ryan Lively) for all their input and support. We would also like to acknowledge our Fundraising Chair JR Johnson and the Workshop Chair Yifu Ding, as well as our sponsors who have provied substantial funds to support the conference and conference attendees.

We are truly excited to host you in Tuscaloosa and look forward to seeing many old friends and meeting new ones. We hope you have a great time at NAMS 2023!

Milad Esfahani, Steven Weinman, and Jason Bara
Chemical & Biological Engineering, Th University of Alabama
NAMS 2023 Conference Co-Chairs

NAMS 2023 Anti-Harassment Policy

NAMS 2023 is dedicated to providing a harassment-free conference experience for everyone. We will not tolerate harassment of conference participants (including sexual harassment and assault) in any form. Conference participants violating these rules may be sanctioned or expelled from the conference without a refund at the discretion of the conference organizers. Harassment includes but is not limited to: comments related to gender, gender identity, and expression, sexual orientation, physical appearance, race, or religion; unwanted sexual attention; unwanted physical contact; threats of violence; intimidation; and stalking.

Reporting: If someone makes you or anyone else feel unsafe or unwelcome, please report it as soon as possible to one of the conference co-chairs. If you are being harassed by someone, notice that someone else is being harassed, or have any other concerns, please contact one the conference co-chairs through the Whova App or in person.

Enforcement: If a participant engages in harassing behavior, event organizers retain the right to take any actions to keep the event a welcoming environment for all participants. This includes warning the offender or expulsion from the conference with no refund. Event organizers may take action to redress anything designed to, or with the clear impact of, disrupting the event or creating a hostile environment for attendees. We expect participants to follow these rules at all event venues and event-related social activities.

NAMS 2023 ALAN S. MICHAELS RECIPIENT

This award is given to honor the late Dr. Michaels and to recognize individuals who have made outstanding innovations and/or exceptional lifetime contributions to membrane science and technology.

Prof. Dibakar Bhattacharyya (DB), University of Kentucky

Dr. Dibakar (DB) Bhattacharyya, UK Alumni Professor of Chemical Engineering at the University of Kentucky, received the Alan S. Michaels Award for his work in the development of stimuli responsive functionalized membranes and the application of membrane technology to solve critical environmental and societal challenges, including the development of high-performance masks and air filtration systems for protection against COVID-19 infection. DB has also been a mentor to literally hundreds of undergraduate students and dozens of PhD students.

In addition, Dr. Bhattacharyya's work has had a direct impact on the commercialization of membrane technologies. He worked with Nanostone Water (now Solecta) on the synthesis of positively-charged nanofiltration membranes that have used to treat wastewater generated from flue gas desulfurization operations. His patents on functionalized membranes have been employed at a Superfund remediation site as part of a process to degrade critical organics like trichloroethylene and perchloroethylene in groundwater samples. More recently, DB's research on COVID-19 has resulted in patent applications: "Antiviral mask and antiviral filter made from a breathable microporous polymeric membrane", and "Hollow fiber membrane for enclosed space air remediation."

The breadth of DB's contributions to membrane science is truly impressive. He has developed novel graphene-based nanofiltration membranes. He has provided fundamental insights into transport mechanisms during pervaporation through zeolite membranes. He has examined the use of porous membranes for the synthesis of bimetallic metal catalysts. He has performed wide-ranging studies of functionalized membranes for the removal of heavy metals from wastewater streams.

DB has also been an active contributor to the membrane community, having organized 3 separate NAMS meetings in 1992, 2001, and 2018 (all in Lexington, KY) while serving as a member of the Board of Directors for many years and as President of NAMS (from 2015-2016). The Alan S. Michaels Award recognizes DB's outstanding innovations and lifetime contributions to membrane science and technology and to the broader membrane community.

NAMS 2023 FELLOWS

NAMS Fellows are recognized members of NAMS who have demonstrated ongoing excellence in membrane engineering and practice. Election as Fellow is in recognition of both “service to NAMS” and “highly significant professional accomplishment in the membrane field.”

Isabel Escobar

Dr. Escobar is a Professor of Chemical and Materials Engineering at the University of Kentucky where she also serves as the Associate Director of the Center of Membrane Sciences. Isabel has held impactful leadership positions within NAMS such as President (2016-2017), Vice President (2015-2016), and Board Member (2005-2020). In addition, she has co-instructed the very important NAMS pre-conference workshop,“Membranes for Water Treatment”, and co-chaired very successful NAMS Annual Meetings in 2007, 2012, and 2017. Her research activities have resulted in over 80 publication, 2 patents, and over 175 conference presentations. She is also a winner of the 2011 American Institute of Chemical Engineers Separations Division FRI/John G. Kunesh Award that recognizes outstanding contributions to the academic, scientific, technological, industrial, or service areas involving separations technologies for individuals under the age of 40.

G. Glenn Lipscomb

Dr. Lipscomb is Distinguished University Professor of Chemical Engineering at the University of Toledo. Glenn was a NAMS Board Member for a decade during which he also held President and Vice-President positions. He was the NAMS Secretary (and Office Manager) for 10 years from 2001-2011, as well as, developer and manager for the NAMS meeting website for 8 annual meetings. Additionally, he co-chaired the NAMS Annual Meeting in 1998 and 2011, as well as, performing duties as Technical Program co-chair in 2007 & '08. He co-chaired the very successful International Congress on Membranes and Membrane Processes in 2017. He has co-instructed the very important NAMS Annual Meeting pre-conference Workshop on "Module design and process calculations/gas separations" 17 times since 2000. Glenn has over 80 peer-reviewed publications, patents, and edited books and is a frequent invited lecturer at international venues, including being University of Bologna Institute of Advanced Studies Fellow in 2015. He is a past Chair of the AIChE Separations Division and was awarded its 2021 Founders Award.

Award Session – Druid, 9:30 AM -12:30 PM, Wednesday, May 17th

Student Fellowship Award

The NAMS Student Fellowship Awards are presented annually to outstanding graduate students in the area of membrane science and technology.

Youngjoo Lee
Georgia Institute of Technology
10:30 AM, Wednesday
Druid

Jishan Wu
University of California, Los Angeles
9:30 AM, Wednesday
Druid

Nicole Moehring
Vanderbilt University
10:00 AM, Wednesday
Druid

Young Membrane Scientist Award

The NAMS Young Membrane Scientist Awards are awarded to those individuals who are within five years of completing their Ph.D. degree and have already proven to be outstanding membrane scientists in the area of membrane science and technology.

Steven T. Weinman
The University of Alabama
11:00 AM, Wednesday
Druid

Lingxiang Zhu
National Energy Technology Laboratory
11:30 am, Wednesday
Druid

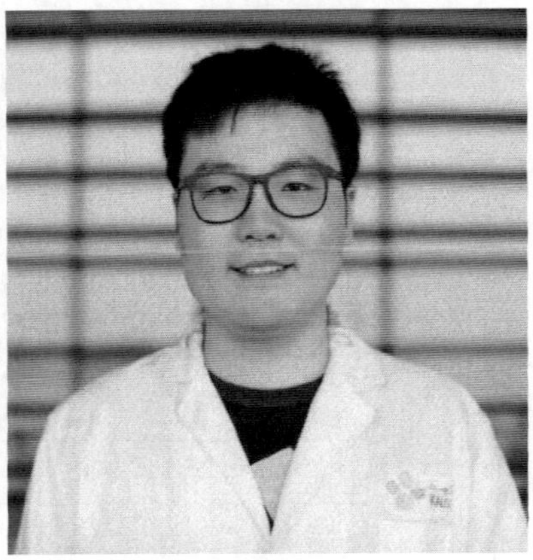

Jie Shen
King Abdullah University of Science and Technology
12:00 PM, Wednesday
Druid

NAMS

The North American Membrane Society

➤ PROGRAM-AT-A-GLANCE

NAMS 2023 Meeting

Date	Time (*CST)	Event	ROOM
Saturday May 13	8:30 AM – 5:00 PM	Workshop I: Membranes for Water Treatment and Reuse	Druid
Sunday May 14	8:00 AM - 5:00 PM	Workshop III: Membranes for Gas Separations	Druid
		Workshop IV: Membranes in Bioprocessing	Capstone
	2: 00 PM - 5:00 PM	NAMS Student Workshop	Yellow Hammer
	7:30 AM – 6:00 PM	Registration	Lobby
	6:00 PM - 9:00 PM	Welcome Reception	Ballroom
Monday May 15	7:00 AM – 5:00 PM	Registration	Lobby
	8:00 AM - 9:00 AM	Plenary Lecture 1	Ballroom
	9:00 AM – 9:30 AM	Coffee Break	Ballroom Foyer
	9:30 AM – 12:30 PM	Parallel Technical Sessions I	See Session Details
	12:30 PM – 2:00 PM	Lunch with Legends	FIVE Tuscaloosa
	2:00 PM - 5:00 PM	Parallel Technical Sessions II	See Session Details
	7:00 PM – 10:00 PM	Poster Session	Ballroom
Tuesday May 16	7:00 AM – 5:00 PM	Registration	Lobby
	8:00 AM – 9:00 AM	Plenary Lecture II	Ballroom
	9:00 AM – 9:30 AM	Coffee Break	Ballroom Foyer
	9:30 AM – 12:30 PM	Parallel Technical Sessions III	See Session Details
	12:30 PM - 2:00 PM	Lunch Break	
	2:00 PM – 5:00 PM	Parallel Technical Sessions IV	See Session Details
	5:30 PM – 6:30 PM	NAMS Business Meeting	Private Dining
	6:30 PM - 7:30 PM	Closing and Award	Ballroom
	7:30 PM - 11:00 PM	Banquet Dinner	River Market
Wednesday May 17	7:00 AM – 5:00 PM	Registration	Lobby
	8:00 AM – 9:00 AM	Plenary Lecture III	Ballroom
	9:00 AM – 9:30 AM	Coffee Break	Ballroom Foyer
	9:30 AM – 12:30 PM	Parallel Technical Sessions V	See Session Details
	12:30 PM – 2:00 PM	Lunch Break	
	2:00 PM – 5:00 PM	Parallel Technical Sessions VI	See Session Details

*CST – US Central Standard time, UK +6 hrs, Germany +7 hrs, Saudi Arabia (in Riyadh) +8 hrs, Inda (IST) +10.5 hrs, Singapore +13 hrs, China +13 hrs, South Korea +14 hrs, Japan +14 hrs.

NAMS
The North American Membrane Society

➢ PROGRAM-AT-A-GLANCE

NAMS 2023 Meeting

Date	Time (*MST)	Event	Room
Monday May 16	8:00 AM - 9:00 AM	**Plenary Session 1**	**Ballroom**
	9:00 AM – 9:30 AM	**Break**	
	9:30 AM – 12:30 PM (Parallel Oral Sessions)	**Polymer Design for Membranes**	**Druid**
		Industrial Membranes 1	**Yellow Hammer**
		Membrane Synthesis and Casting 1	**Black Warrior**
		Carbon Capture and Facilitated Transport	**Capstone**
		Module Modeling and Design	**Private Dining**
	12:30 PM – 2:00 PM	**Lunch Break**	
	2:00 PM - 5:00 PM (Parallel Oral Sessions)	**Inorganic Membrane Materials**	**Druid**
		Industrial Membranes 2	**Yellow Hammer**
		Catalytic and Responsive Membranes	**Black Warrior**
		Electrochemical Separations 1	**Capstone**
		Desalination and Osmotic Processes	**Private Dining**
	7:00 PM – 10:00 PM	**Poster Session**	**Ballroom**
Tuesday May 17	8:00 AM – 9:00 AM	**Plenary Session 2**	**Ballroom**
	9:00 AM – 9:30 AM	**Break**	
	9:30 AM – 12:30 PM (Parallel Oral Sessions)	**Polymeric and Mixed-Materials – Gas Separations**	**Druid**
		Downstream Bioprocessing	**Yellow Hammer**
		Membrane Synthesis and Casting 2	**Black Warrior**
		Highly Selective Separations	**Capstone**
		Emerging Materials for Liquid Separations	**Private Dining**
	12:30 PM - 2:00 PM	**Lunch Break**	
	2:00 PM – 5:00 PM (Parallel Oral Sessions)	**Molecular and Process Modeling**	**Druid**
		Organic Solvent Separations	**Yellow Hammer**
		Membrane Characterization	**Black Warrior**
		Electrochemical Separations 2	**Capstone**
		Membrane Distillation and Pervaporation	**Private Dining**
Wednesday May 18	8:00 AM – 9:00 AM	**Plenary Session 3**	**Ballroom**
	9:00 AM – 9:30 AM	**Break**	
	9:30 AM – 12:30 PM (Parallel Oral Sessions)	**NAMS Awards Session**	**Druid**
		Membranes for Gas Separations	**Yellow Hammer**
		Contaminant Removal from Water	**Black Warrior**
		Membrane Fouling	**Capstone**
		Process Intensification and Integration	**Private Dining**
	12:30 PM – 2:00 PM	**Lunch Break**	
	2:00 PM – 5:00 PM (Parallel Oral Sessions)	**High Salinity Streams and Electrochem 3**	**Druid**
		Water Reuse	**Yellow Hammer**
		Bioinspired and Biomimetic Membranes	**Black Warrior**
		Membrane and Module Optimization	**Capstone**
			Private Dining

*CST – US Central Standard time, UK +6 hrs, Germany +7 hrs, Saudi Arabia (in Riyadh) +8 hrs, Inda (IST) +10.5 hrs, Singapore +13 hrs, China +13 hrs, South Korea +14 hrs, Japan +14 hrs.

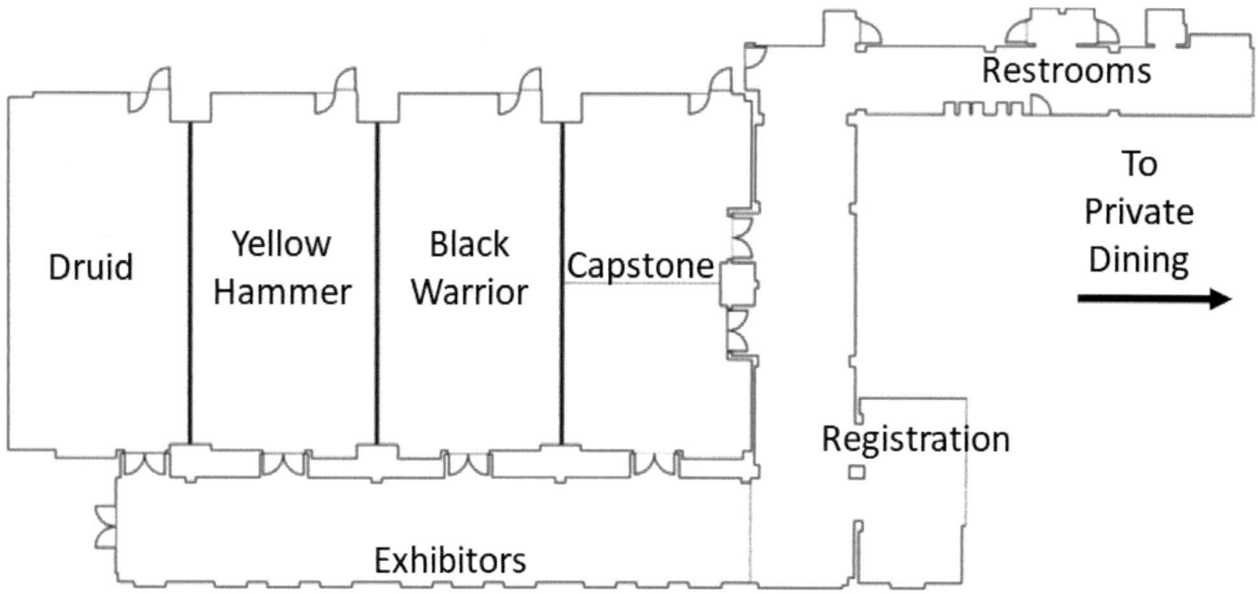
Restrooms
Druid
Yellow Hammer
Black Warrior
Capstone
To Private Dining
Registration
Exhibitors

8:00 AM	Plenary Session 1 (Ballroom)				
Parallel Sessions	**Room: Druid** Polymer Design for Membranes	**Room: Yellow Hammer** Industrial Membranes 1	**Room: Black Warrior** Membrane Synthesis and Casting 1	**Room: Capstone** Carbon Capture and Facilitated Transport	**Room: Private Dining** Module Modeling and Design
Chairs	**Priyesh Wagh** priyesh.wagh@lubrizol.com **Alexander Lopez** amlopez2@olemiss.edu **Devin Shaffer** dshaffer@uh.edu	**Dibakar Bhattacharyya** db@uky.edu **CJ Kurth** cj.kurth@gmail.com **Christina Carbrello** christina.carbrello@milliporesigma.com	**Xiaobo Dong** xiaobo_dong@pall.com **Reza Foudazi** rfoudazi@ou.edu **Abhishek Roy** abhishek.roy@nrel.gov	**Katherine Hornbostel** hornbostel@pitt.edu **David Hopkinson** David.Hopkinson@netl.doe.gov **Chen Zhang** czhang71@umd.edu **Casey O'Brien** cobrie23@nd.edu	**David Ladner** ladner@clemson.edu **Grigorios Panagakis** gpanagak@andrew.cmu.edu
9:30 AM	Oral 1 - Versatile synthetic platform using click chemistry to study water and ion transport in post-functionalized membranes Joshua D Moon (University of Florida)	Oral 7 - An Improved Electrodialysis Method for Amino Acid Production Jian Xu (Bayer Crop Science)	Oral 13 - Functional Hollow Fibers Made from Polymers of Intrinsic Microporosity (PIMs) for Membrane-Based Separations Philippe Jean-Baptiste (Massachusetts Institute of Technology)	Oral 25 - Scale-up of CO_2-selective nanoporous single-layer graphene membranes Jian Hao (Ecole polytechnique fédérale de Lausanne)	Oral 19 - Modeling gas separation in flat sheet membrane modules: Impact of flow channel size variation Joanna Rivero (University of Pittsburgh)
10:00 AM	Oral 2 - Deep Eutectic Solvent (DES) – Inspired Design of Membranes for CO_2/Gas Separation Narjes Esmaeili (The State University of New York at Buffalo)	Oral 8 - Some Key Considerations for Developing New Industrial Membranes John F Thompson (NL Chemical Technology)	Oral 14 - Carbon Capture Membranes Based on Ultrathin Amorphous Polyethers Enabled by Nanoconfinement and Interfacial Engineering Haiqing Lin (The State University of New York at Buffalo)	Oral 26 - Extending the effects of crosslinks during free volume manipulation (FVM) of amine-functionalized PIM-1 membranes for gas separation Taigyu Joo (Massachusetts Institute of Technology)	Oral 20 - Hollow Fiber Membrane Modules for Flue Gas Separation: Experimental Verification of CFD Models using 3D Printing Victor A Kusuma (NETL)
10:30 AM	Oral 3 - High performance gas separation membranes from CANAL ladder polymers Ashley Robinson (Stanford University)	Oral 9 - Repairing the Membrane Innovation Pipeline: How to Reconnect Academics and Industry For more Impactful Membrane Research Jeffrey R McCutcheon (University of Connecticut)	Oral 15 - Utilizing Greener Solvents for Membrane Fabrication Joseph T Danner (The University of Alabama)	Oral 27 - Graphene Oxide/Single Walled Carbon Nanotube (GO-SWCNT) Network Restricted Ionic Liquid as Stable and Effective Membrane for Highly Efficient Carbon Capture Dinesh K Behera (University at Buffalo)	Oral 21 - Harnessing gravity to mitigate polarization and increase permeate production in membrane distillation Nils Tilton (Colorado School of Mines)
11:00 AM	Oral 4 – Upcycled Polyvinyl Chloride (PVC) Electrospun Nanofibers from Waste PVC based Materials for Water Treatment Atta Ur Razzaq (The University of Alabama)	Oral 10 - Designing a New Generation of Reverse Osmosis Membranes with Chemical Resistance Sue J Mecham (NALA Membranes)	Oral 16 - Combining Macrostructure, Microstructure and Molecular Scale Morphology to Create New High-Performance Composite Carbon Molecular Sieve Membranes Yuhe Cao (Georgia Institute of Technology)	Oral 28 - Effect of Amine Content on the Carbon Dioxide Separation Performance of Polyvinylamine Membranes for Flue Gas Carbon Capture James S Baker (National Energy Technology Laboratory)	Oral 22 - Demonstration of Direct Ocean Carbon Capture Using Membrane Technologies Joanna Rivero (University of Pittsburgh)
11:30 AM	Oral 5 - Highly Permselective Contorted Polyamide Desalination Membranes Devin Shaffer (University of Houston)	Oral 11 - Scale-up and commercialization of a novel membrane chemistry Christopher J Drover (ZwitterCo, Inc.)	Oral 17 - Highly selective interfacial polymerized polyamide TFC membranes by tunning synthesis parameters Chenhao Yao (University of Texas at Austin)	Oral 29 - Carbon capture from flue gases using an integrated membrane skid at ~1 tonne CO_2/day scale Yang Han (The Ohio State University)	Oral 23 - Energy and exergy analysis of multi-stage vacuum membrane distillation integrated with mechanical vapor compression Mahdi Malmali (Texas Tech University)
12:00 PM	Oral 6 - Understanding the Structure-Property Relationship of Poly(Ether Ether Ketone)-Ionene Composite Membranes for Gas Separations Sudhir Ravula (University of Alabama)	Oral 12 - 3D-Printed Spacer Membrane Elements CJ Kurth (Aqua Membranes)	Oral 18 - Electrospray Printed Membranes for Gas Separations Noah Ferguson (University of Connecticut)	Oral 30 - Hydrogen sulfide and acetylene stability and the effect of $AgTf_2N$ concentration on fractional free volume of hydrogen-stable Ag^+-based facilitated transport membranes Matthew N Davenport (The University of Texas at Austin)	Oral 24 - A new approach to simulating concentration polarization in reverse osmosis systems with feed spacers Nils Tilton (Colorado School of Mines)

Monday Afternoon

Parallel Sessions	Room: Druid Inorganic Membrane Materials	Room: Yellow Hammer Industrial Membranes 2	Room: Black Warrior Catalytic and Responsive Membranes	Room: Capstone Electrochemical Separations 1	Room: Private Dining Desalination and Osmotic Processes
Chairs	**Olivia Lenz** olivia@membrion.com **Yang Lui** yang.liu@mtrinc.com	**Dibakar Bhattacharyya** db@uky.edu **CJ Kurth** cj.kurth@gmail.com **Christina Carbrello** christina.carbrello@milliporesigma.com	**Ayse Asatekin** ayse.asatekin@tufts.edu **Simona Liguori** sliguori@clarkson.edu	**Hee Jeung Oh** hjoh@psu.edu **Shudipto Dishari** sdishari2@unl.edu **Piran Kidambi** piran.kidambi@vanderbilt.edu **William Tarpeh** wtarpeh@stanford.edu	**Andrea Achilli** achilli@arizona.edu **Jonathan Maisonneuve** maisonneuve@oakland.edu **William Phillip** wphillip@nd.edu
2:00 PM	Oral 31 - Aramid-derived Carbon Molecular Sieve Membranes for Carbon Dioxide Capture Gaurav Iyer (University of Maryland)	Oral 37 - Development of MTR Polaris Membranes for Carbon Capture Tim Merkel (Membrane Technology & Research Inc.)	Oral 43 - Thermoresponsive antifouling ultrafiltration membranes from mesophase templating Reza Foudazi (University of Oklahoma)	Oral 49 - Advanced electrochemical membranes for capacitive deionization Christopher G Arges (The Pennsylvania State University)	Oral 55 - Membrane Design Criteria and Practical Viability of Pressure-driven Distillation Weifan Liu (Vanderbilt University)
2:30 PM	Oral 32 - Interfacial Polymerization of Nanoparticles for Modifiable, Ultra-Thin, Ultrafiltration Membranes Dean F Stipanic (University of Toronto)	Oral 38 - Novel Catalyst-Coated Membranes for Green H2 Production Chunqing Liu (Honeywell UOP)	Oral 44 - Engineering a biocatalytic membrane for caffeine upcycling Hadi Rouhi (University of Alabama)	Oral 50 - Ion Containing Membranes For Clean Energy and Sustainable Environment Abhishek Roy (NREL)	Oral 56 - Nitrate and Salt Transport Coefficients Determination for Spiral-Wound RO Membranes Derived from Field Data of Treatment of Impaired Brackish Groundwater Yoram Cohen (University of California, Los Angeles)
3:00 PM	Oral 33 - Balanced reduction and Oxidation to Engineer Graphene Oxide Membranes for Dye Desalination Haiqing Lin (The State University of New York at Buffalo)	Oral 39 - Timeline of Membrane Development David Hasse (Air Liquide)	Oral 45 - Reverse Osmosis Silica Fouling Control using Reactive Micromixing Ruizhe Xu (University of Texas at Austin)	Oral 51 - Bio-Inspired Approaches to Manage Ion Transport Limitation and Directionality of Ion Transport in Electrochemical Systems Shudipto K Dishari (University of Nebraska-Lincoln)	Oral 57 - Effectively Using Heat to Thermally Enhance Osmotic Processes for Power Generation and Desalination Sanjana Chintalacheruvu (Oakland University)
3:30 PM	Oral 34 - Pyridinic nitrogen substituted two-dimensional pores on single-layer graphene for rapid and selective CO_2 transport Kuang-Jung Hsu (EPFL)	Oral 40 - Parametric Analysis of Membrane Properties on Post Combustion Flue Gas System Economics Christine Parrish (Compact Membrane Systems)	Oral 46 - Composite Hollow Fiber Membrane Reactor Containing Immobilized Organocatalysts and Palladium for Sustainable Chemical Transformation Ali A Rownaghi (U.S. Department of Energy/Cleveland State University)	Oral 52 - Impact of hierarchical cation-exchange membranes' chemistry and crosslinking level on electrodialysis demineralization performances of a complex food solution Elodie Khetsomphou (Laval University)	Oral 58 - Effect of Dehydration on RO Desalination Performance in Fully Aromatic Polyamide Membranes Containing TEMPO-Oxidized Cellulose Nanocrystals (TOCNs) Connor L Farrell (Virginia Tech)
4:00 PM	Oral 35 - Carbon molecular sieve membranes derived from molecularly engineered cross-linkable 6FDA-based copolyimides for gas separations Zhongyun Liu (Georgia Institute of Technology)	Oral 41 - Energy Efficient Olefin-Paraffin Separations with Optiperm™ Membrane System Brandon J Burghardt (Compact Membrane Systems)	Oral 47 - Conversion of waste biomass to platform chemicals and fuels using a catalytic membrane reactor Ranil Wickramasinghe (University of Arkansas)	Oral 53 - Factors Influencing Ion Transport in Activated Wafer Electrodeionization Leticia Santos de Souza (University of Arkansas)	Oral 59 - Reverse osmosis and membrane distillation: potential tools for leveraging surplus energy at nuclear power plants for seawater desalination and societal benefit Laura H Arias Chavez (Tennessee Tech University)
4:30 PM	Oral 36 - Ultrasonication Assisted Membrane Surface Functionalization by Silver (Ag)- and Zinc (ZIF-8) Metal-Organic Frameworks for Water Treatment: Synthesis, Characterization, and Performance Himangshu Mondal (The University of Alabama)	Oral 42 - Enabling a Robust Hydrogen Economy via Advanced MEA and Membranes Noah D Meeks (Southern Company Services, Inc.)	Oral 48 - Effect of PPE decontamination processes on porous polysulfone (PSf) flat sheet membranes fabricated via nonsolvent-induced phase separation (NIPS): A comparison with traditional PPE Ebuka Ogbuoji (University of Kentucky)	Oral 113 - Ion-exchange biosponge polymers for selective removal of toxic ionic chemotherapy drugs Hee Jeung Oh (Pennsylvania State University)	Oral 60 - Pressure-induced diffusion of solvents in dense homogeneous membranes Kevin Reimund (University of Texas at Austin)
7:00 PM	PPoste6Sessior(Ballroom)				

8:00 AM	Plenary Session 2 (Ballroom)				
Parallel Sessions	**Room: Druid** Polymeric and Mixed Materials for Gas Separations	**Room: Yellow Hammer** Downstream Bioprocessing	**Room: Black Warrior** Membrane Synthesis and Casting 2	**Room: Capstone** Highly Selective Separations	**Room: Private Dining** Emerging Materials for Liquid Separations
Chairs	**Benjamin Sundell** benjaminjsundell@gmail.com **Zachary Smith** zpsmith@mit.edu **Xiaoli Ma** ma26@uwm.edu **Raj Singh** rsingh@lanl.gov	**Onur Kas** Onur.Kas@ucb.com **Zhao Li** zhao.li3@merck.com	**Xiaobo Dong** xiaobo_dong@pall.com **Reza Foudazi** rfoudazi@ou.edu **Abhishek Roy** abhishek.roy@nrel.gov	**Christine Duval** ced84@case.edu **Syed Islam** islamsz@ornl.gov **Joshua Moon** joshua.moon@ufl.edu	**Oishi Sanyal** oishi.sanyal@mail.wvu.edu **Volodymyr Tarabara** tarabara@egr.msu.edu
9:30 AM	Oral 85 - Long term stability of thin film cross-linked ionic liquid membranes for CO_2 gas separation Richard D. Noble (University of Colorado, Boulder)	Oral 67 - Characterization and Associated Pressure-Dependent Behavior of Deposited mRNA-Lipid Nanoparticles Formed During Sterile Filtration Kevork Oliver Messerian (The Pennsylvania State University)	Oral 73 - Towards better understanding of synthesis-morphology-performance in polyamide membranes: Visualization under controlled reaction conditions Adi Ben Zvi (Technion)	Oral 79 - Specific Ion Selectivity in Sulfonated Polystyrene Membranes Near the Percolation Threshold Yuxuan Huang (Columbia University)	Oral 61 - Solvent-Free Carrier-Based Membranes for Selective Ion Separations Jay Werber (University of Toronto)
10:00 AM	Oral 86 - Rational design of mixed-matrix metal-organic framework membranes for molecular separations Shuvo Jit Datta (KAUST)	Oral 68 - Fouling of Virus Clearance Membranes Yair Kaufman (TeraPore Technologies)	Oral 74 - Molecularly mixed composite membranes based on triptycene-isatin Porous Polymer Networks and Carboxylic-acid Functionalized PIM-1 Lucas Condes (University of Oklahoma)	Oral 80 - Chelating DGA membranes enable ion-ion selectivity in radiopharmaceutical processing Christine E Duval (Case Western Reserve University)	Oral 62 - Structure and transport studies of carbon molecular sieve membranes for wastewater treatment Young Hee Yoon (Georgia Institute of Technology)
10:30 AM	Oral 87 - Highly selective polymer membranes containing poly(amic-acid) Stabilized Silver Nanoparticles Matthew T Webb (University of Oklahoma)	Oral 69 - Surrogate-based Optimization for the Techno-economic Feasibility Analysis of Membrane Capture Chromatography Platforms Juan I Romero (Clemson University)	Oral 75 - Functionalized NF and adsorptive membranes for the separation of perfluoro-organics and environmentally relevant ions Francisco Leniz (university of Kentucky)	Oral 81 - Selective extraction of medium-chain fatty acid from organic waste streams using supported liquid membranes Jongho Lee (University of British Columbia)	Oral 63 - Differences in water and vapor transport through angstrom-scale pores in atomically thin membranes Piran Kidambi (Vanderbilt University)
11:00 AM	Oral 88 - Tailoring Ultramicropores in Hybrid Carbon Molecular Sieve Membranes for H_2 Purification Leiqing Hu (University at Buffalo)	Oral 70 - Valorization of Food Byproducts inside Processing Industries Shanti Bhushan (GEA)	Oral 76 - Lignin-based membrane fabricated with deep eutectic solvent Abaynesh Gebreyohannes (King Abdullah University of Science and Technology)	Oral 82 - Foundational Framework for Evaluating Performance of Nanofiltration based Solute-solute Separation Shihong Lin (Vanderbilt University)	Oral 64 - Removal of Micro- and Nano-plastics from Water using Liquid-Liquid Extraction Ashish Srivastava (The University of Alabama)
11:30 AM	Oral 89 - Analysis of gas transport in molecularly-mixed composite membranes Matthew Rivera (MIT)	Oral 71 - Virus adhesion to archetypal membrane surfaces: Developing guidance on membrane selection for virus removal Vlad Tarabara (Michigan State)	Oral 77 - Scaling-Up and Modelling the Fabrication of a More Sustainable Polymeric Membrane for Water Treatment Using Slot Die Coating David Lu (University of Kentucky)	Oral 83 - A Novel Membrane Solvent Extraction Process Enabling Highly Efficient Separation and Recovery of Critical Materials from End-of-Life Lithium-Ion Batteries Shailesh Singh Dangwal (Oak Ridge National Lab)	Oral 65 - Synthesis and fabrication of mesoporous membranes with enhanced mechanical properties from amphiphilic multiblock polymers via self-assembly and nonsolvent-induced phase separation Ben Pedretti (University of Texas at Austin)
12:00 PM	Oral 90 - Interfacial poly(epoxyether) membranes, a versatile thin-film synthesis platform for carbon capture membranes Daan Van Havere (KU Leuven)	Oral 72 - Improved Filtration Performance of Virus Filters at High Product Titers Ranil Wickramasinghe (University of Arkansas)	Oral 78 - Patterning Super-oleophobic Poly (vinylidene fluoride) Membranes via Confined Thermally Induced Phase Separation for Enhanced Oil-Water Separation Shouhong Fan (University of Colorado at Boulder)	Oral 84 - Isolation of Rare Earth Ions Using Electrodialysis or Nanofiltration Merlin Bruening (University of Notre Dame)	Oral 66 – NONE

Tuesday Afternoon

Parallel Sessions	Room: Druid Molecular and Process Modeling	Room: Yellow Hammer Organic Solvent Separations	Room: Black Warrior Membrane Characterization	Room: Capstone Electrochemical Separations 2	Room: Private Dining Membrane Distillation and Pervaporation
Chairs	**Shihong Lin** shihong.lin@vanderbilt.edu **Akshay Deshmukh** akshaypd@mit.edu	**Michele Galizia** mgalizia@ou.edu **Neel Rangnekar** neel.d.rangnekar@exxonmobil.com	**Abhishek Mondal** abhishek.mondal@donaldson.com **Sudhir Ravula** sravula@ua.edu	**Hee Jeung Oh** hjoh@psu.edu **Shudipto Dishari** sdishari2@unl.edu **Piran Kidambi** piran.kidambi@vanderbilt.edu **William Tarpeh** wtarpeh@stanford.edu	**Lucy Mar Camacho** Lucy.Camacho@tamuk.edu **Tiezheng Tong** Tiezheng.Tong@colostate.edu
2:00 PM	Oral 91 - Molecular Dynamics Simulations of Charge Interactions in Polyamide Membranes Richard Lueptow (Northwestern University)	Oral 97 - Transport Property Modulation via Non-Aqueous Solvent Specific Behavior in Cross-Linked Membranes Charles Leroux (University of Virginia)	Oral 103 - Improved Mechanical Compliance in Bicontinuous Cubic Lyotropic Membranes Based on a Cross-Linking Gemini Monomer via Copolymerization with a Non-Cross-Linkable Analog Richard D. Noble (University of Colorado, Boulder)	Oral 109 - A new polymer platform for anion exchange membranes based on Friedel-Crafts polymerization Jennifer L Schaefer (University of Notre Dame)	Oral 115 - Anti-scalants for mitigating silica scaling in membrane desalination: Effects of molecular structure and membrane process Yiqun Yao (Colorado State University)
2:30 PM	Oral 92 - Water Model Choice for Molecular Dynamics Simulations of Polyamide Membranes Suwei Liu (Northwestern University)	Oral 98 - Engineering organic solvent reverse osmosis performance of vapor phase infiltrated organic-inorganic hybrid membranes Yi Ren (Georgia Institute of Technology)	Oral 104 - Advanced metrology, molecular dynamics, and machine learning for determining ion activity coefficients in ion-exchange membranes Christopher G Arges (The Pennsylvania State University)	Oral 110 - Highly Charged Ion-Exchange Membranes for Treating Concentrated Salt Solutions via Electrodialysis Jovan Kamcev (University of Michigan)	Oral 116 - New approaches to mitigate wetting and scaling in membrane distillation Shihong Lin (Vanderbilt University)
3:00 PM	Oral 93 - Physical mechanism of water transport in reverse osmosis membranes: Solution-diffusion or pore-flow? Li Wang (Tongji University)	Oral 99 - Development of a high-performance 30nm NIPS PTFE membrane Karl Ashkar (Pall Corporation)	Oral 105 - Mathematical Modeling of CO_2 Facilitated Transport across Amine-functionalized Polymeric Membranes with Direct Operando Observation of Amine Carrier Saturation Casey O'Brien (University of Notre Dame)	Oral 111 - Production of polyphenols-enriched cranberry juice through electrodialysis with filtration membrane: How physicochemical properties of filtration membranes impact anthocyanins migration and selectivity during electrodialysis Eva Revellat (Université Laval)	Oral 117 - Hydrophobic Hollow Fiber-Membrane for Spacesuit Cooling Dibakar Bhattacharyya (University of Kentucky)
3:30 PM	Oral 94 - Machine Learning Modeling of Intermittent Reverse Osmosis Desalination System Operation via Transfer Learning and Long Short-term Memory Yoram Cohen (University of California, Los Angeles)	Oral 100 - Crosslinked thin film composite membranes for industrial organic complex mixtures separations Yacine Feliachi (Georgia Tech)	Oral 106 - Unveiling the mechanism of light gas transport through configurational free volume in glassy polymers William Box (University of Oklahoma)	Oral 112 - Correlation between volatile fatty acid transport properties in anion exchange membranes and volatile fatty acid transport during electrodialysis Holly M Haflich (University of North Carolina- Chapel Hill)	Oral 118 - Radiofrequency responsive membranes for membrane distillation applications Mahdi Malmali (Texas Tech University)
4:00 PM	Oral 95 - Impact of dynamical correlations on salt transport in solvated ion exchange membranes Nico Marioni (The University of Texas at Austin)	Oral 101 - Ultrathin polytriazole membranes for energy-intensive separation Stefan Chisca (KAUST)	Oral 107 - Characterizing dielectric properties and water molecule dynamics using dielectric relaxation spectroscopy Sean M Bannon (University of Virginia)	Oral 54 - Bipolar membrane electrodialysis for direct LiOH production from concentrated brines Aubrey Quigley (The University of Texas at Austin)	Oral 119 - Materials and design to enhance transport in membrane distillation Waqas Alam (Purdue University)
4:30 PM	Oral 96 - Monovalent Ion Selectivity of Cellulose Acetate Membranes Paul R Irving (University of Texas at Austin)	Oral 102 - Unveiling the role of sorption, diffusion, interactions and plasticization in organic solvent transport in dense OSN/OSRO membranes Michele Galizia (University of Oklahoma)	Oral 108 - A Comprehensive OMICs Platform for Detecting Biofouling in Full-Scale Reverse Osmosis Facilitates Weiliang Bai (University of Texas at Austin)	Oral 114 - Revolutionizing Ion Transport: Dynamic Channel-Size Tuning in MXene Transistor Membranes for Selective Separation and Enhanced Control Aaditya A Pendse (Lawrence Livermore National Labortory)	Oral 120 - Omniphobic electrospun membranes for treatment of hydraulic fracturing produced water through membrane distillation Ranil Wickramasinghe (University of Arkansas)

Wednesday Morning

8:00 AM	Plenary Session 3 (Ballroom)				
Parallel Sessions	**Room: Druid** NAMS Awards	**Room: Yellow Hammer** Membranes for Gas Separations	**Room: Black Warrior** Contaminant Removal from Water	**Room: Capstone** Membrane Fouling	**Room: Private Dining** Process Intensification and Integration
Chairs	**Ryan Lively** ryan.lively@chbe.gatech.edu **Evan Hatakeyama** EHatakeyama@chevron.com	**Ali Rownaghi** a.rownaghi@csuohio.edu **Surya Tiwari** surya.tiwari@netl.doe.gov	**Jung Min Kim** xeu8wp@virginia.edu **Boya Xiong** bxiong@umn.edu	**Daniel Miller** dan.miller@twelve.co **Audie Thompson** audie.k.thompson@usace.army.mil **Ngoc Bui** ngoctbui21@ou.edu	**Marie-Eve Langevin** MELangevin@ameridia.com **Mahdi Malmali** mahdi.malmali@ttu.edu **Albert Kim** AlbertSK@hawaii.edu **Jia Wei Chew** JChew@ntu.edu.sg
9:30 AM	Oral 121 - Challenges and opportunities in ultra-high pressure reverse osmosis operation Jishan Wu (UCLA)	Oral 145 - Co^{2+}-based supramolecular networks achieving extraordinary CO_2/O_2 selectivity Taliehsadat Alebrahim (The State University of New York at Buffalo)	Oral 133 - Polyamide Reverse Osmosis Membrane Modification with Diamines to Improve Small Neutral Molecule Rejection Shahriar Habib (The University of Alabama)	Oral 139 - Energy reduction in ceramic microfiltration using supersaturated carbon-dioxide-enhanced backwash Weiming Qi (Clemson University)	Oral 127 - Thermally and Chemically Stable Metallic Membrane Reactors for H_2 Generation with CO_2 Capture Hani A.E. Hawa (Precision Combustion, Inc.)
10:00 AM	Oral 122 - Kinetic Control of Angstrom-scale Porosity in Graphene for Direct Scalable Synthesis of Atomically Thin Proton Exchange Membranes Nicole K Moehring (Vanderbilt University)	Oral 146 - Crown Ether-Based Polyimides for Sour Gas Separation Dana Wong (Aramco Americas)	Oral 134 - Electrocatalytic laser-induced graphene membrane for micropollutant removal and biofouling control Swatantra Pratap Singh (IIT Bombay)	Oral 140 - The role of surfactants in membrane fouling Naama Segev-Mark (Technion)	Oral 128 - Specific Energy Analysis of Using Fertilizer-Based Liquid Desiccants to Dehumidify Indoor Plant Environments Sarah Moussaddy (Oakland University)
10:30 AM	Oral 123 - Data-driven predictions of complex mixture permeation in polymer membranes Youngjoo Lee (Georgia Tech)	Oral 147 - Modeling Gas Permeation in Amorphous Polymer Membranes Using Non-Equilibrium Molecular Dynamics Simulations Mohammed N Al Otmi (University of Florida)	Oral 135 - Simultaneous Uranium Extraction and Brine Desalination via Adsorptive Pervaporation by A Bifunctional Graphene Oxide Membrane Guangcheng Wang (National University of Singapore)	Oral 141 - In-situ measurements to differentiate homogeneous and heterogeneous nucleation in membrane distillation: effects of spacer geometry and antiscalants Mahdi Malmali (Texas Tech University)	Oral 129 - Feasibility of Membrane-adsorption Hybrid Process for Direct Air Capture Yi-Chen Huang (The Ohio State University)
11:00 AM	Oral 124 - Modification and Synthesis of Polyamide Reverse Osmosis Membranes for the Separation of Small, Neutral Molecules Steven T Weinman (The University of Alabama)	Oral 148 - Selective H_2S/CO_2 Separation using Sterically Hindered Amine Membranes Shraavya Rao (Ohio State University)	Oral 136 - Simple and low-cost electroactive membranes for ammonia recovery Xinyi Wang (University of California, Los Angeles)	Oral 142 - Antimicrobial Properties of Chitosan-Graphene Oxide Composites Christopher S Griggs (U.S. Army ERDC)	Oral 130 - Electrospray 3D Printing to Unlock Customized Membrane Performance Mert Can Hacifazlioglu (University of Connecticut)
11:30 AM	Oral 125 - High-Performance Non-Aging Rubbery Thin Film Composite Membranes for Post-Combustion CO_2 Capture: From Material Synthesis to Field Demonstration Lingxiang Zhu (National Energy Technology Laboratory)	Oral 149 - Sweetening Sour Gas with Hybrid Membranes Patrick T Wright (Aramco Americas)	Oral 137 - MIL-100 (Fe) Functionalized Thin-film Nanocomposite Membranes for Enhanced Removal of Uncharged Urea from Water Tin Le (The University of Alabama)	Oral 143 - Spacers for Mass Transfer Enhancement in Flat Sheet Membrane Modules – An Upper Bound? Glenn Lipscomb (University of Toledo)	Oral 131 - Intensified membrane distillation for treating oilfield produced waters from unconventional wells: Energetics analysis and pilot demonstration Mahdi Malmali (Texas Tech University)
12:00 PM	Oral 126 - Tuning 2D materials channels for sub-nano scale separation: from interlayers to in-plane pores Jie Shen (King Abdullah University of Science and Technology)	Oral 150 - Identifying key factors to unlock plasticization resistance in microporous polymer membranes Aristotle F Grosz (MIT)	Oral 138 - Microcystin-LR destruction and removal from water via enzymatic biodegradation and ultrafiltration Abelline K Fionah (University of Kentucky)	Oral 144 - Real-Time Induced Magnetic Vibrational Based Antifouling Mechanism for Ultrafiltration (UF) membrane Jasneet K Pala (The University of Alabama)	Oral 132 - Low concentration carbon dioxide capture using MEEP-based membranes Birendra Adhikari (Idaho National Laboratory)

Wednesday Afternoon

Parallel Sessions	Room: Druid High Salinity Streams and Electrochemical Separations 3	Room: Yellow Hammer Water Reuse	Room: Black Warrior Bioinspired and Biomimetic Membranes	Room: Capstone Membrane and Module Optimization	Room: Private Dining
Chairs	**Kofi Christie** kchristie@lsu.edu **Bianca Souza Chaves** bchaves@arizona.edu **Jonathan Brant** jbrant1@uwyo.edu	**Sage Hiibel** shiibel@unr.edu **Vicky Karanikola** vkaranik@email.arizona.edu	**Cassandra Porter** cjp0084@auburn.edu **Yuexiao Shen** Yuexiao.Shen@ttu.edu **Harekrushna Behera** chem.harekrushna@gmail.com	**Hani Hawa** hhawa@precision-combustion.com	N/A
2:00 PM	Oral 151 - Sulfonated Polystyrene Membranes Near the Percolation Threshold for Hypersaline Electrodialysis Desalination Yuxuan Huang (Columbia University)	Oral 157 - Advancements in Water Treatment: Concentrating RO Residuals Contaminated with PFAS through Closed Circuit RO and Membrane Distillation Tae Lee (US EPA)	Oral 163 - Tunable membranes incorporating artificial water channels for high-performance water reverse osmosis desalination Mihail Barboiu (Institut Europeen des Membranes)	Oral 169 - Random polyampholyte amphiphilic copolymers for membranes with tunable selectivity and excellent fouling resistance Ayse Asatekin (Tufts University)	N/A
2:30 PM	Oral 152 - Controlled Localized Metal Organic Framework Synthesis on Anion Exchange Membranes Harm Wiegerinck (University of Twente)	Oral 158 - Backwashable hollow-fibre membranes enable ultrahigh recovery nanofiltration process for effluent reuse Chidiebere S Nnebuo (Ben-Gurion University of the Negev)	Oral 164 - Development of Nanofiltration membranes incorporating Artificial Water Channel Omar Samhari (Université de Montpellier)	Oral 170 - Co-transport of methanol and formate in crosslinked phenyl acrylate-based ion exchange membranes: Effect of steric hindrance Jung Min Kim (University of Virginia)	
3:00 PM	Oral 153 - Assessing the Suitability of Nanofiltration for Pretreatment of Lithium-Enriched Brines Juan Zhai (Texas Tech University)	Oral 159 - A membrane-based multi-barrier approach to potable water reuse and waste heat recovery Andrea Achilli (University of Arizona)	Oral 165 - Effect of Nanopatterning on E. Coli Fouling and Concentration Polarization Lauren M Ward (The University of Alabama)	Oral 171 - Acoustic actuated 3D printed bubble-trapping structures for in-situ mitigation of membrane fouling Kieran R Fung (University of Colorado Boulder)	
3:30 PM	Oral 154 - Ion transport in ion exchange membranes: water confinement effects at low hydration José Carlos Díaz (University of Michigan)	Oral 160 - Transport mechanisms of fouled engineering-scale water reuse reverse osmosis membranes Bianca M. Souza Chaves (The University of Arizona)	Oral 166 - Biomimetic Supramolecular Channels for Sustainable Extraction of Rare Earth Elements Harekrushna Behera (The University of Texas at Austin)	Oral 172 - High-performance large-area polymeric membranes with chlorine tolerance for water purification by pressure-driven distillation Duong T. Nguyen (University of Colorado Boulder)	
4:00 PM	Oral 155 - Extending the reverse osmosis to osmotic mediation for brine management Ali Naderi Beni (Purdue University)	Oral 161 - Selective removal of selenium from saline water using different MOFs and MOF-incorporated NF membrane Sweta A Modak (The University of Alabama)	Oral 167 - Biomimetic membranes with high water vapor permeance and protective capacity for next generation breathable protective fabrics Hyeonji Oh (University of Texas at Austin)	Oral 173 - Graphene Oxide and Sulfonated graphene oxide embedded Polyamide-polysulfone based Thin-film Nanocomposite Sea water Reverse Osmosis membranes Asim K Ghosh (Bhabha Atomic Research Centre)	
4:30 PM	Oral 156 - Understanding mixed salt partitioning in model poly(ethylene oxide)-based membrane materials Everett S Zofchak (The University of Texas at Austin)	Oral 162 – NONE	Oral 168 – NONE	Oral 174 - Sweeping Gas membrane distillation for rapid heat dissipation Waqas Alam (Purdue University)	

Ballroom, 8:00 am - 9:00 am, Monday, May 15th

Membrane Science in Support of Moon to Mars: Challenges and Opportunities

Plenary Speaker

Dr. James F. Reilly

Retired NASA astronaut, Honorary United States Marshal, and Former Director of the United States Geological Survey

Dr. Jim Reilly is an Executive Advisor in Booz Allen's aerospace business. As a space systems expert and former astronaut, Jim works with the Booz Allen teams to develop innovative solutions for space-related services, earth systems characterization, and customer-facing applications for a variety of clients including the Department of Defense, National Oceanic and Atmospheric Administration, National Aeronautics and Space Administration (NASA), and the intelligence community. Jim is also active in supporting the development of In-space Servicing, Assembly, Manufacturing, Mobility and Logistics (ISAMML) capabilities with a collaboration of 7 major universities, NASA, and industry. Objectives are to reduce the cost of in-space and surface systems for Earth, Moon, and Mars programs by moving away from traditional terrestrial processes to methods operating in situ. Jim has more than 25 years of multidisciplinary technical, engineering, and management experience in resource assessments; applications-focused earth systems characterization; space systems development, applications, and operations; and education. Prior to joining Booz Allen, Jim Reilly was the Director of the U.S. Geological Survey (USGS) where he managed an annual $1.8 billion operating budget for the nation's largest water, earth resources, biological science, natural hazards, and civilian mapping agency. Before USGS, Jim served U.S. and allied militaries as a subject matter expert on space operations, and as a technical advisor supporting the National Security Space Institute of the U.S. Air Force Space Command. As an astronaut at NASA, he had a distinguished 13-year career where he flew on three space shuttle missions and conducted five spacewalks totaling more than 856 hours in space.In addition to Senior Management positions at TAEUS Corporation and PhotoStencil Corporation in Colorado Springs, Colorado, Jim has held management positions in the private sector, academia, in government service, in the military, and has owned his own space technology-focused company. He served 17 years with the U.S. Navy as a reserve engineering duty officer and was designated a space systems expert in the Navy's Space Cadre. Jim earned his bachelor's, master's, and doctorate degrees in geosciences from the University of Texas at Dallas.

Abstract

Long duration human exploration of the Moon and Mars will require a change in the concept of operations regarding crew life support and mission environmental controls. The International Space Station has operated regenerable life support systems in an operational test environments for over a decade. Recent upgrades to the water recycling system have included a membrane Brine Processing Assembly to increase recoveries above the 85% recovery ratio experienced with the previous system. Atmospheric regeneration has been reliant on oxygen replenishment from Earth and to a smaller extent on electrolytic decomposition of water. For long duration missions away from Earth, water enroute and in support of flora/fauna as nutritional supplies, will need to be maintained within closed loop systems. On the surface water may be sourced from in situ resources in the form of ices on the Moon and in subsurface ice sources on Mars. Water recovery and treatment (Mars waters are likely to contain significant perchlorate concentrations based on surface analyses) will be required to provide metabolic and atmospheric make up supplies. Membrane treatment systems provide low cost/mass/power options for required processing/reprocessing of this resource. Further, should active agriculture be an objective for Mars habitats, the Mars soils will need to be treated, again with high efficiency membrane systems.

Experimental and exploration-related applications will develop the scalable systems for Earth based remediation benefitting disadvantaged populations across the globe.

Ballroom, 8:00 am - 9:00 am, Tuesday, May 16th

Imparting biofunctional properties to nanoconjugates and nanostructured membranes for advanced separation, conversion, and sensing systems

Plenary Speaker

Dr. Lidietta Giorno

National Research Council of Italy, Institute on Membrane Technology

Dr. Lidietta Giorno has served as the director of the Institute on Membrane Technology of the National Research Council (ITM-CNR) of Italy since 2009. She is an expert in membrane science and membrane bioengineering with a particular focus on nanostructured biohybrid membranes, biocatalytic membrane reactors, membrane emulsification, integrated membrane processes for water, biotechnology, pharmaceutical, and biorefinery. She was awarded the International Awards of "Guido Dorso" for research in 2011, sponsored by the Italian Senate and the University of Naples Federico II. She was awarded the "Sapio Red Carpet" Award in 2016 among the female scientists of the highest scientific profile who are engaged in the development of the Country. Dr. Giorno is the co-author of 13 books, over 200 peer-reviewed papers, and co-editor of the Encyclopedia of Membranes, Springer, 2016. She has served on the European Membrane Society (EMS) Council as the President. She is an Honorary Member of the European Membrane Society since 2014.

Abstract

The perspective of circular economy, sustainable and responsible production requires highly precise, selective, flexible and efficient processes. Artificial membranes and membrane-based operations play a pivotal role in this scenario. On the other hand, their precision, selectivity and flexibility are far from those expressed by natural membranes. Indeed, biological system membranes exhibit the unpaired capability to selectively recognize, transport, transform, autoregulate, self-clean, self-heal, regenerate, etc. Although biomolecules are generally labile, their orderly, hierarchical and intensified assembly results in system structures that can operate efficiently for several decades (i.e., a lifetime). This suggests that artificial membranes could perform much better than the current state-of-the-art. Understanding the structures, mechanisms, and systems by which biological system membranes function and operate may inspire new strategies in the development of advanced materials, membranes, and modes of operation. The lecture will survey some of the biostructures and fluid dynamics strategies used in natural systems, highlighting similarities and differences to lab-made membranes and modes of operation. For example, one of the key aspects by which the membranes of biological systems achieve high selectivity is permeation through multilevel gates from larger pores to smaller pores (i.e., the opposite of what occurs in man-made membranes, where permeation occurs from a dense or microporous thin layer to a porous thick support). Concentration gradient (or concentration polarization if ions are involved) and fouling are prevented and controlled by applying a low transmembrane pressure and thus a low permeate flux ($L^3 \cdot t^{-1} \cdot L^{-2}$). High productivity (i.e., flow rate, $L^3 \cdot t^{-1}$) in a small footprint "module" is achieved by packing a large membrane area into an extremely intensified system (surface-to-volume ratio $> 1 \cdot 10^5$ m^2/m^3).

One of the strategies to improve the performance of synthetic membranes is to impart them biological functions. This is particularly important in the areas of bio-related production, separation, remediation, sensing, and detection, and has the potential to be extended to the chemical and energy fields. The lecture will highlight recent results on combining biomacromolecules with highly selective recognition properties (such as enzymes) with nanostructured membranes. Recent observations suggest that membranes can assist the formation of bio-nanoconjugates in which the enzyme can be accommodated with a conformation that is most active and selective. The topography and flexibility of bio-nanoconjugates play a critical role in preserving selectivity, activity, and stability. The ability of membranes to accommodate different biomacromolecules in different compartments opens new scenarios for intensifying precise and flexible processes and reducing byproducts. Case studies demonstrating the potential of such biofunctional hybrid membranes for bioactive molecule production (including optically pure enantiomers), bioremediation, and biosensing will be discussed. In particular, the impressive attention that biofunctionalized membranes are attracting in biosensors for environmental monitoring will be illustrated.

Ballroom, 8:00 am - 9:00 am, Wednesday, May 17th

The Energy Transition and Its Opportunities

Plenary Speaker

Dr. Noah D. Meeks, P.E.

Principal Research Engineer with Southern Company Services R&D

Noah D. Meeks, Ph.D., P.E. is a Principal Research Engineer with Southern Company Services R&D, working in the Advanced Energy Systems Research Group for the past seven years. His primary focus is technology development for the hydrogen economy and its potential in economy-wide decarbonization. The portfolio of technologies under development include those for hydrogen production, infrastructure, and efficient energy utilization. Previously, Noah was a research engineer in the Environmental Controls Research Group at Southern, having joined the Company in 2012. Noah holds a Ph.D. in chemical engineering from the University of Kentucky, where his dissertation under Prof. DB Bhattacharyya focused on development of functionalized silica for environmental applications. He has a B.S. in chemistry from Northern Kentucky University. He is a Professional Engineer in Alabama. Noah and his wife, Elaine, live in Hoover, AL, along with their twin toddlers Samuel and Joshua.

Abstract

The Energy Transition describes moving a global energy economy to net-zero energy carriers without compromising standards of living. Electricity, transportation fuel, thermal energy, and feedstocks must all transition to net zero, which represents a massive shift toward zero-carbon primary energy and net zero carbon feedstocks. In addition, the energy stocks that ensure the smooth functioning of the global economy must be maintained in energy-dense media. All of this must be done in a way that protects the environment, food supply, and developing economies. Once the scale of this challenge is understood, only a few dominant technological solutions are available. Nuclear energy must be rapidly deployed not only in support of current electricity needs, but in support of future electrification, industrial energy, and hydrogen. Generating the baseload, abundant, and carbon-free energy will require significant technological development and its availability may even alter the technologies of choice for the global economy.

Oral Sessions - Monday

Presenting authors are indicated by an asterisk*.

Oral 1 – Polymer Design for Membranes

9:30 AM – 12:30 PM, Druid

Session Chairs

Priyesh Wagh
Lubrizol Corporation

Alexander Lopez
University of Mississippi

Devin Shaffer
University of Houston

9:30 AM (Druid, Monday)
Oral 1 - Versatile synthetic platform using click chemistry to study water and ion transport in post-functionalized membranes
Joshua D Moon (University of Florida)*; Oscar Nordness (University of California, Santa Barbara); Peter Richardson (University of California, Santa Barbara); Rahul Sujanani (University of California, Santa Barbara); Raphaële Clément (University of California, Santa Barbara); Benny Freeman (The University of Texas at Austin); Rachel Segalman (University of California, Santa Barbara); Craig Hawker (University of California, Santa Barbara)

10:00 AM (Druid, Monday)
Oral 2 - Deep Eutectic Solvent (DES) – Inspired Design of Membranes for CO2/Gas Separation
Narjes Esmaeili (The State University of New York at Buffalo)*; Haiqing Lin (The State University of New York at Buffalo)

10:30 AM (Druid, Monday)
Oral 3 - High performance gas separation membranes from CANAL ladder polymers
Ashley Robinson (Stanford University)*

11:00 AM (Druid, Monday)
Oral 4 - Upcycled Polyvinyl Chloride (PVC) Electrospun Nanofibers from Waste PVC-based Materials for Water Treatment
Atta Ur Razzaq (The University of Alabama)*; Milad Rabbani Esfahani (The University of Alabama)

11:30 AM (Druid, Monday)
Oral 5 - Highly Permselective Contorted Polyamide Desalination Membranes
Devin Shaffer (University of Houston)*

12:00 PM (Druid, Monday)
Oral 6 - Understanding the Structure-Property Relationship of Poly(Ether Ether Ketone)-Ionene Composite Membranes for Gas Separations
Sudhir Ravula (University of Alabama)*; Kevin W Wise (University of Alabama); Jason Bara (University of Alabama)

Oral 2 – Industrial Membranes I

9:30 AM – 12:30 PM, Yellow Hammer

Session Chairs

Dibakar Bhattacharyya
University of Kentucky

CJ Kurth
Aqua Membranes

Christina Carbrello
MilliporeSigma

9:30 AM (Yellow Hammer, Monday)
Oral 7 - An Improved Electrodialysis Method for Amino Acid Production
Jian Xu (Bayer Crop Science)*

10:00 AM (Yellow Hammer, Monday)
Oral 8 - Some Key Considerations for Developing New Industrial Membranes
John F Thompson (NL Chemical Technology)*; Norman Li (NL Chemical Technology)

10:30 AM (Yellow Hammer, Monday)
Oral 9 - Repairing the Membrane Innovation Pipeline: How to Reconnect Academics and Industry For more Impactful Membrane Research
Jeffrey R McCutcheon (University of Connecticut)*; Edward Wazer (University of Connecticut); Marianne Pemberton (University of Connecticut); Mayur Ostwal (University of Connecticut); Mi Zhang (University of Connecticut); Yara Suleiman@uconn.edu (University of Connecticut); Sina Shahbazmohamadi (University of Connecticut)

11:00 AM (Yellow Hammer, Monday)
Oral 10 - Designing a New Generation of Reverse Osmosis Membranes with Chemical

Resistance
Sue J Mecham (NALA Membranes)*; Judy Riffle (NALA Membranes); Mikayla Armstrong (NALA Membranes); Andrew Tibbits (NALA Membranes)

11:30 AM (Yellow Hammer, Monday)
Oral 11 - Scale-up and commercialization of a novel membrane chemistry
Christopher J Drover (ZwitterCo, Inc.)*

12:00 PM (Yellow Hammer, Monday)
Oral 12 - 3D-Printed Spacer Membrane Elements
CJ Kurth (Aqua Membranes)*

Oral 3 – Membrane Synthesis and Casting 1

9:30 AM – 12:30 PM, Black Warrior

Session Chairs
Xiaobo Dong
Pall Corporation

Reza Foudazi
University of Oklahoma

Abhishek Roy
National Renewable Energy Laboratory

9:30 AM (Black Warrior, Monday)
Oral 13 - Functional Hollow Fibers Made from Polymers of Intrinsic Microporosity (PIMs) for Membrane-Based Separations
Philippe Jean-Baptiste (Massachusetts Institute of Technology)*; Zachary Smith (MIT)

10:00 AM (Black Warrior, Monday)
Oral 14 - Carbon Capture Membranes Based on Ultrathin Amorphous Polyethers Enabled by Nanoconfinement and Interfacial Engineering
Haiqing Lin (The State University of New York at Buffalo)*; Gengyi Zhang (The State University of New York at Buffalo); Vinh Bui (University at Buffalo)

10:30 AM (Black Warrior, Monday)
Oral 15 - Utilizing Greener Solvents for Membrane Fabrication
Joseph T Danner (The University of Alabama)*; Steven T Weinman (The University of Alabama)

11:00 AM (Black Warrior, Monday)
Oral 16 - Combining Macrostructure, Microstructure and Molecular Scale Morphology to Create New High-Performance Composite Carbon Molecular Sieve Membranes
Yuhe Cao (Georgia Institute of Technology)*; Zhongyun Liu (Georgia Institute of Technology); Wulin Qiu (Georgia Institute of technology); William Koros (Georgia Institute of Technology)

11:30 AM (Black Warrior, Monday)
Oral 17 - Highly selective interfacial polymerized polyamide TFC membranes by tunning synthesis parameters
Chenhao Yao (University of Texas at Austin)*; Manish Kumar (University of Texas at Austin)

12:00 PM (Black Warrior, Monday)
Oral 18 - Electrospray Printed Membranes for Gas Separations
Noah Ferguson (University of Connecticut)*; Srivatsa Bettahalli (University of Connecticut); Mayur Ostwal (University of Connecticut); Jeffrey R McCutcheon (University of Connecticut)

Oral 4 – Module Modeling and Design

9:30 AM – 12:30 PM, Private Dining

Session Chairs
David Ladner
Clemson University

Grigorios Panagakis
Carnegie Melon University

9:30 AM (Private Dining, Monday)
Oral 19 - Modeling gas separation in flat sheet membrane modules: Impact of flow channel size variation
Joanna Rivero (University of Pittsburgh)*; Glenn Lipscomb (University of Toledo, Toledo); Katherine Hornbostel (University of Pittsburgh); Leo Nemetz (University of Toledo); Marcos Da Conceicao Acosta (University of Toledo)

10:00 AM (Private Dining, Monday)
Oral 20 - Hollow Fiber Membrane Modules for Flue Gas Separation: Experimental Verification of CFD Models using 3D Printing
Victor A Kusuma (NETL)*; Lili Sun (University of Toledo); Comfort Oluleke (University of Toledo); Nathan Diemler (NETL); Lingxiang Zhu (National Energy Technology Laboratory); Grigorios Panagakos

(Carnegie Mellon University); Glenn Lipscomb (University of Toledo, Toledo); David Hopkinson (NETL)

10:30 AM (Private Dining, Monday)
Oral 21 - Harnessing gravity to mitigate polarization and increase permeate production in membrane distillation
Nils Tilton (Colorado School of Mines)*; Federico Municchi (Colorado School of Mines); Yiming Liu (UCLA); Tzahi Cath (Colorado School of Mines); Michael Heeley (Colorado School of Mines); Eric Hoek (UCLA); Craig S Turchi (NREL); David Jassby (UCLA)

11:00 AM (Private Dining, Monday)
Oral 22 - Demonstration of Direct Ocean Carbon Capture Using Membrane Technologies
Joanna Rivero (University of Pittsburgh)*; Austin R Lieber (University of Pittsburgh); Chris Snodgrass (University of Pittsburgh); Zoe Neal (University of Pittsburgh); Donna Hildebrandt (University of Pittsburgh); William Gamble (University of Pittsburgh); Shanna Davidson (University of Pittsburgh); Huda Usman (University of Pittsburgh); Tagbo Niepa (University of Pittsburgh)

11:30 AM (Private Dining, Monday)
Oral 23 - Energy and exergy analysis of multi-stage vacuum membrane distillation integrated with mechanical vapor compression
Mahdi Malmali (Texas Tech University)*; Bosong Lin (Texas Tech University)

12:00 PM (Private Dining, Monday)
Oral 24 - A new approach to simulating concentration polarization in reverse osmosis systems with feed spacers
Nils Tilton (Colorado School of Mines)*; Jacob Johnston (Colorado School of Mines); Sarah Dischinger (LBNL); Mostafa Nassr (UT Austin); Ji Yeon Li (LBNL); Sergi Molins (Lawrence Berkeley National Laboratory); Nicolas Spycher (LBNL); Benny Freeman (University of Texas Austin); William Stringfellow (LBNL); Daniel Miller (LBNL); Kristofer Gleason (University of Texas Austin)

Oral 5 – Carbon Capture and Facilitated Transport

9:30 AM – 12:30 PM, Capstone

Session Chairs
Katherine Hornbostel
University of Pittsburgh

David Hopkinson
National Energy Technology Laboratory

Chen Zhang
University of Maryland

Casey O'Brien
University Of Notre Dame

9:30 AM (Capstone, Monday)
Oral 25 - Scale-up of CO_2-selective nanoporous single-layer graphene membranes
Jian Hao (Ecole polytechnique fédérale de Lausanne)*; Kumar Varoon Agrawal (Ecole polytechnique fédérale de Lausanne)

10:00 AM (Capstone, Monday)
Oral 26 - Extending the effects of crosslinks during free volume manipulation (FVM) of amine-functionalized PIM-1 membranes for gas separation
Taigyu Joo (Massachusetts Institute of Technology)*; Sungsool Wi (National High Magnetic Field Laboratory); Zachary Smith (MIT)

10:30 AM (Capstone, Monday)
Oral 27 - Graphene Oxide/Single Walled Carbon Nanotube (GO-SWCNT) Network Restricted Ionic Liquid as Stable and Effective Membrane for Highly Efficient Carbon Capture
Dinesh K Behera (University at Buffalo)*; Bratin Sengupta (University at Buffalo); Fan Wang (University at Buffalo); Shiguang Li (Gas Technology Institute); Miao Yu (University at Buffalo)

11:00 AM (Capstone, Monday)
Oral 28 - Effect of Amine Content on the Carbon Dioxide Separation Performance of Polyvinylamine Membranes for Flue Gas Carbon Capture
James S Baker (National Energy Technology Laboratory)*; Victor A Kusuma (NETL); Zi Tong (National Energy Technology Laboratory); David Hopkinson (NETL)

11:30 AM (Capstone, Monday)

Oral 29 - Carbon capture from flue gases using an integrated membrane skid at ~1 tonne CO_2/day scale

Yang Han (The Ohio State University)*; Yutong Yang (The Ohio State University); Babul Prasad (The Ohio State University); Ruizhi Pang (The Ohio State University); Winston Ho (OSU)

12:00 PM (Capstone, Monday)

Oral 30 - Hydrogen sulfide and acetylene stability and the effect of $AgTf_2N$ concentration on fractional free volume of hydrogen-stable Ag^+-based facilitated transport membranes

Matthew N Davenport (The University of Texas at Austin)*; Tangqiumei Song (The University of Texas at Austin); Caitlin Bentley (The University of Texas at Austin); Taylor McClung (The University of Texas at Austin); Cara Doherty (CSIRO); Joan Brennecke (The University of Texas at Austin); Benny Freeman (The University of Texas at Austin)

Oral Sessions - Monday

Presenting authors are indicated by an asterisk*.

Oral 6 – Inorganic Membrane Materials

2:00 PM – 5:00 PM, Druid

Session Chairs

Olivia Lenz
Membrion

Yang Liu
Membrane Technology & Research, Inc

2:00 PM (Druid, Monday)
Oral 31 - Aramid-derived Carbon Molecular Sieve Membranes for Carbon Dioxide Capture
Gaurav Iyer (University of Maryland)*; Chen Zhang (University of Maryland)

2:30 PM (Druid, Monday)
Oral 32 - Interfacial Polymerization of Nanoparticles for Modifiable, Ultra-Thin, Ultrafiltration Membranes
Dean F Stipanic (University of Toronto)*

3:00 PM (Druid, Monday)
Oral 33 - Balanced reduction and Oxidation to Engineer Graphene Oxide Membranes for Dye Desalination
Haiqing Lin (The State University of New York at Buffalo)*; Erda Deng (State University of New York at Buffalo); Kai Chen (University at Buffalo); Mengqi Yuan (University at Buffalo)

3:30 PM (Druid, Monday)
Oral 34 - Pyridinic nitrogen substituted two-dimensional pores on single-layer graphene for rapid and selective CO_2 transport
Kuang-Jung Hsu (EPFL)*

4:00 PM (Druid, Monday)
Oral 35 - Carbon molecular sieve membranes derived from molecularly engineered cross-linkable 6FDA-based copolyimides for gas separations
Zhongyun Liu (Georgia Institute of Technology)*; William Koros (Georgia Institute of Technology)

4:30 PM (Druid, Monday)
Oral 36 - Ultrasonication Assisted Membrane Surface Functionalization by Silver (Ag)- and Zinc (ZIF-8) Metal-Organic Frameworks for Water Treatment: Synthesis, Characterization, and Performance
Himangshu Mondal (The University of Alabama)*; Medha Kasula (The University of Alabama); Milad Rabbani Esfahani (The University of Alabama)

Oral 7 – Industrial Membranes 2

2:00 PM – 5:00 PM, Yellow Hammer

Session Chairs

Dibakar Bhattacharyya
University of Kentucky

CJ Kurth
Aqua Membranes

Christina Carbrello
MilliporeSigma

2:00 PM (Yellow Hammer, Monday)
Oral 37 - Development of MTR Polaris Membranes for Carbon Capture
Tim Merkel (Membrane Technology & Research Inc.)*; Johannes G Wijmans (Membrane Technology and Research, Inc.)

2:30 PM (Yellow Hammer, Monday)
Oral 38 - Novel Catalyst-Coated Membranes for Green H_2 Production
Chunqing Liu (Honeywell UOP)*

3:00 PM (Yellow Hammer, Monday)
Oral 39 - Timeline of Membrane Development
David Hasse (Air Liquide)*; Sandeep Karode (Air Liquide); Marcos Da Conceicao (Air Liquide)

3:30 PM (Yellow Hammer, Monday)
Oral 40 - Parametric Analysis of Membrane Properties on Post Combustion Flue Gas System Economics
Joshua C Watson (Compact Membrane Systems, Inc.); Christine Parrish (Compact Membrane Systems)*; Ken Loprete (Compact Membrane Systems)

4:00 PM (Yellow Hammer, Monday)
Oral 41 - Energy Efficient Olefin-Paraffin Separations with Optiperm™ Membrane System
Brandon J Burghardt (Compact Membrane

Systems)*; Christine Parrish (Compact Membrane Systems); Ken Loprete (Compact Membrane Systems)

4:30 PM (Yellow Hammer, Monday)
Oral 42 - Enabling a Robust Hydrogen Economy via Advanced MEA and Membranes
Noah D Meeks (Southern Company Services, Inc.)*

Oral 8 – Catalytic and Responsive Membranes

2:00 PM – 5:00 PM, Black Warrior

Session Chairs
Ayse Asatekin
Tufts University

Simona Liguori
Clarkson University

2:00 PM (Black Warrior, Monday)
Oral 43 - Thermoresponsive antifouling ultrafiltration membranes from mesophase templating
Saadatgharehbagh Younes (University of Oklahoma); Seyed Mostafa Tabatabaei (University of Oklahoma); Kyungtae Kim (Los Alamos National Laboratory); Reza Foudazi (University of Oklahoma)*

2:30 PM (Black Warrior, Monday)
Oral 44 - Engineering a biocatalytic membrane for caffeine upcycling
Hadi Rouhi (University of Alabama)*; Meredith Mock (University of Alabama); Ryan Summers (The University of Alabama); Steven T Weinman (The University of Alabama)

3:00 PM (Black Warrior, Monday)
Oral 45 - Reverse Osmosis Silica Fouling Control using Reactive Micromixing
Ruizhe Xu (the University of Texas at Austin)*; Weiliang Bai (University of Texas at Austin); Manish Kumar (University of Texas at Austin)

3:30 PM (Black Warrior, Monday)
Oral 46 - Composite Hollow Fiber Membrane Reactor Containing Immobilized Organocatalysts and Palladium for Sustainable Chemical Transformation
Ali A Rownaghi (U.S. Department of Energy/Cleveland State University)*

4:00 PM (Black Warrior, Monday)
Oral 47 - Conversion of waste biomass to platform chemicals and fuels using a catalytic membrane reactor
Ranil Wickramasinghe (University of Arkansas)*; Xianghong Qian (University of Arkansas); Chhabilal Regmi (University of Arkansas); Zhexi Zhu (University of Arkansas)

4:30 PM (Black Warrior, Monday)
Oral 48 - Effect of PPE decontamination processes on porous polysulfone (PSf) flat sheet membranes fabricated via nonsolvent-induced phase separation (NIPS): A comparison with traditional PPE
Ebuka Ogbuoji (University of Kentucky)*; Isabel Escobar (University of Kentucky)

Oral 9 – Electrochemical Separations 1

2:00 PM – 5:00 PM, Capstone

Session Chairs
Hee Jeung Oh
Penn State University

Shudipto Dishari
University of Nebraska-Lincoln

Piran Kidambi
Vanderbilt University

William Tarpeh
Stanford University

2:00 PM (Capstone, Monday)
Oral 49 - Advanced electrochemical membranes for capacitive deionization
Christopher G Arges (The Pennsylvania State University)*; Tanmay Kulkarni (The Pennsylvania State University); Bharat Shrimant (The Pennsylvania State University)

2:30 PM (Capstone, Monday)
Oral 50 - Ion Containing Membranes For Clean Energy and Sustainable Environment
Abhishek Roy (NREL)*

3:00 PM (Capstone, Monday)
Oral 51 - Bio-Inspired Approaches to Manage Ion Transport Limitation and Directionality of Ion Transport in Electrochemical Systems
Shudipto K Dishari (University of Nebraska-Lincoln)*

3:30 pm (Capstone, Monday)
Oral 52 - Impact of hierarchical cation-exchange membranes' chemistry and crosslinking level on electrodialysis

demineralization performances of a complex food solution
Elodie Khetsomphou (Laval University, Institute of Nutrition and Functional Food (INAF))*; Laurent Bazinet (Laval University, Institute on Nutrition and Functional Food (INAF))

4:00 pm (Capstone, Monday)
Oral 53 - Factors Influencing Ion Transport in Activated Wafer Electrodeionization
Leticia Santos de Souza (University of Arkansas)*

4:30 pm (Capstone, Monday)
Oral 113 - Ion-exchange biosponge polymers for selective removal of toxic ionic chemotherapy drugs
Hee Jeung Oh (Pennsylvania State University)*

Oral 10 – Desalination and Osmotic Processes

2:00 PM – 5:00 PM, Private Dining

Session Chairs
Andrea Achilli
University of Arizona

Jonathan Maisonneuve
Oakland Univeristy

William Phillip
University of Notre Dame

2:00 PM (Private Dining, Monday)
Oral 55 - Membrane Design Criteria and Practical Viability of Pressure-driven Distillation
Weifan Liu (Vanderbilt University)*; Shihong Lin (Vanderbilt University)

2:30 PM (Private Dining, Monday)
Oral 56 - Nitrate and Salt Transport Coefficients Determination for Spiral-Wound RO Membranes Derived from Field Data of Treatment of Impaired Brackish Groundwater
Nora Marki (University of California, Los Angeles); Yang Zhou (Shanghai University); Kwan Lam Yip (University of California, Los Angeles); Christian Aguilar (University of California, Los Angeles); Yakubu Jarma (University of California, Los Angeles); Bilal Khan (California State University, San Bernardino); Yoram Cohen (UCLA)*

3:00 PM (Private Dining, Monday)
Oral 57 - Effectively Using Heat to Thermally Enhance Osmotic Processes for Power Generation and Desalination
Sanjana Chintalacheruvu (Oakland University)*; Saber Khanmohammadi (Oakland University); Jonathan Maisonneuve (Oakland University)

3:30 PM (Private Dining, Monday)
Oral 58 - Effect of Dehydration on RO Desalination Performance in Fully Aromatic Polyamide Membranes Containing TEMPO-Oxidized Cellulose Nanocrystals (TOCNs)
Connor L Farrell (Virginia Tech)*; E. Johan Foster (University of British Columbia); Stephen M Martin (Virginia Tech)

4:00 PM (Private Dining, Monday)
Oral 59 - Reverse osmosis and membrane distillation: potential tools for leveraging surplus energy at nuclear power plants for seawater desalination and societal benefit
J. Hunter Himes (Tennessee Tech University); Laura H Arias Chavez (Tennessee Tech University)*

4:30 PM (Private Dining, Monday)
Oral 60 - Pressure-induced diffusion of solvents in dense homogeneous membranes
Kevin Reimund (University of Texas at Austin)*; Rahul Sujanani (University of California, Santa Barbara); Jorge Hernandez (University of Texas at Austin); Kristofer Gleason (University of Texas at Austin); Manish Kumar (University of Texas at Austin); Benny Freeman (The University of Texas at Austin)

Oral Sessions - Tuesday

Presenting authors are indicated by an asterisk*.

Oral 11 – Polymeric and Mixed Matrix Materials for Gas Separations

9:30 AM – 12:30 PM, Druid

Session Chairs

Benjamin J. Sundell
C2Sense, Inc

Zachary Smith
Massachusetts Institute of Technology

Xiaoli Ma
University of Wisconsin-Milwaukee

Raj Singh
Los Alamos National Lab

9:30 AM (Druid, Tuesday)
Oral 85 - Long term stability of thin film cross-linked ionic liquid membranes for CO_2 gas separation
Richard D. Noble (University of Colorado, Boulder)*; Douglas Gin (University of Colorado, Boulder); Chamaal Karunaweera (University of Colorado, Boulder); Matthew Cowan (University of Canterbury)

10:00 AM (Druid, Tuesday)
Oral 86 - Rational design of mixed-matrix metal-organic framework membranes for molecular separations
Shuvo Jit Datta (KAUST)*

10:30 AM (Druid, Tuesday)
Oral 87 - Highly selective polymer membranes containing poly(amic-acid) Stabilized Silver Nanoparticles
Matthew T Webb (University of Oklahoma)*; Lucas Condes (University of Oklahoma); Michele Galizia (University of Oklahoma); Sepideh Razavi (University of Oklahoma)

11:00 AM (Druid, Tuesday)
Oral 88 - Tailoring Ultramicropores in Hybrid Carbon Molecular Sieve Membranes for H_2 Purification
Leiqing Hu (University at Buffalo)*; Won Il Lee (Stony Brook University); Soumyabrata Roy (Rice University); Ashwanth Subramanian (Stony Brook University); Kim Kisslinger (Brookhaven National Laboratory); Lingxiang Zhu (National Energy Technology Laboratory); Vinh Bui (University at Buffalo); Shouhong Fan (University of Colorado); Thien N Tran (University at Buffalo, The State University of New York); Gengyi Zhang (The State University of New York at Buffalo); Yifu Ding (Membrane Science, Engineering, and Technology (MAST) Center, Department of Mechanical Engineering, University of Colorado Boulder); Pulickel M. Ajayan (Rice University); Chang-Yong Nam (Brookhaven National Laboratory); Haiqing Lin (The State University of New York at Buffalo)

11:30 AM (Druid, Tuesday)
Oral 89 - Analysis of gas transport in molecularly-mixed composite membranes
Matthew Rivera (MIT)*; Ryan Lively (Georgia Tech)

12:00 PM (Druid, Tuesday)
Oral 90 - Interfacial poly(epoxyether) membranes, a versatile thin-film synthesis platform for carbon capture membranes
Daan Van Havere (KU Leuven)*; Rhea Verbeke (KU Leuven); Benny Freeman (The University of Texas at Austin); Ivo Vankelecom (KU Leuven)

Oral 12 – Downstream Bioprocessing

9:30 AM – 12:30 PM, Yellow Hammer

Session Chairs

Onur Kas
UCB

Zhao Li
Merck

9:30 AM (Yellow Hammer, Tuesday)
Oral 67 - Characterization and Associated Pressure-Dependent Behavior of Deposited mRNA-Lipid Nanoparticles Formed During Sterile Filtration
Kevork Oliver Messerian (The Pennsylvania State University)*

10:00 AM (Yellow Hammer, Tuesday)
Oral 68 - Fouling of Virus Clearance Membranes
Yair Kaufman (TeraPore Technologies)*; Rachel Dorin (Terapore); Gabriel Hale (TeraPore Technologies); KC Hunt (TeraPore Technologies); Daniele Gerion (TeraPore Technologies)

10:30 AM (Yellow Hammer, Tuesday)
Oral 69 - Surrogate-based Optimization for the Techno-economic Feasibility Analysis of Membrane Capture Chromatography Platforms
Juan J Romero (Clemson University)*; Scott Husson (Clemson University); Eleanor Jenkins (Clemson University)

11:00 AM (Yellow Hammer, Tuesday)
Oral 70 - Valorization of Food Byproducts inside Processing Industries
Shanti Bhushan (GEA)*

11:30 AM (Yellow Hammer, Tuesday)
Oral 71 - Virus adhesion to archetypal membrane surfaces: Developing guidance on membrane selection for virus removal
Vlad Tarabara (Michigan State)*; Xunhao Wang (Michigan State University)

12:00 PM (Yellow Hammer, Tuesday)
Oral 72 - Improved Filtration Performance of Virus Filters at High Product Titers
Rony Horax (University of Arkansas); Wenbo XU (University of Arkansas); Xianghong Qian (University of Arkansas); Ranil Wickramasinghe (University of Arkansas)*

Oral 13 – Membrane Synthesis and Casting 2

9:30 AM – 12:30 PM, Black Warrior

Session Chairs
Xiaobo Dong
Pall Corporation

Reza Foudazi
University of Oklahoma

Abhishek Roy
National Renewable Energy Laboratory

9:30 AM (Black Warrior, Tuesday)
Oral 73 - Towards better understanding of synthesis-morphology-performance in polyamide membranes: Visualization under controlled reaction conditions
Adi Ben Zvi (Technion)*; Guy Z Ramon (Technion - Israel Institute of Technology)

10:00 AM (Black Warrior, Tuesday)
Oral 74 - Molecularly mixed composite membranes based on triptycene-isatin Porous Polymer Networks and Carboxylic-acid Functionalized PIM-1
Lucas Condes (University of Oklahoma)*; Matthew T Webb (University of Oklahoma); Adriana Landry (University of Oklahoma); Michele Galizia (University of Oklahoma)

10:30 AM (Black Warrior, Tuesday)
Oral 75 - Functionalized NF and adsorptive membranes for the separation of perfluoro-organics and environmentally relevant ions
Francisco Leniz (University of Kentucky)*; Dibakar Bhattacharyya (University of Kentucky)

11:00 AM (Black Warrior, Tuesday)
Oral 76 - Lignin-based membrane fabricated with deep eutectic solvent
Abaynesh Gebreyohannes (King Abdullah University of Science and Technology (KAUST))*; Suzana Nunes (King Abdullah University of Science and Technology (KAUST))

11:30 AM (Black Warrior, Tuesday)
Oral 77 - Scaling-Up and Modelling the Fabrication of a More Sustainable Polymeric Membrane for Water Treatment Using Slot Die Coating
David Lu (University of Kentucky)*; Kwangjun Jung (Georgia Institute of Technology); Tequila Harris (Georgia Institute of Technology); Isabel Escobar (University of Kentucky)

12:00 PM (Black Warrior, Tuesday)
Oral 78 - Patterning Super-oleophobic Poly (vinylidene fluoride) Membranes via Confined Thermally Induced Phase Separation for Enhanced Oil-Water Separation
Shouhong Fan (University of Colorado at Boulder)*; Duong T. Nguyen (University of Colorado Boulder); Kieran R Fung (Membrane Science, Engineering, and Technology (MAST) Center, Department of Mechanical Engineering, University of Colorado Boulder); Anthony Straub (University of Colorado Boulder); Yifu Ding (Membrane Science, Engineering, and Technology (MAST) Center, Department of Mechanical Engineering, University of Colorado Boulder)

Oral 14 – Highly Selective Separations

9:30 AM – 12:30 PM, Capstone

Session Chairs

Christine Duval
Case Western Reserve University

Syed Islam
Oak Ridge National Laboratory

Joshua Moon
University of Florida

9:30 AM (Capstone, Tuesday)
Oral 79 - Specific Ion Selectivity in Sulfonated Polystyrene Membranes Near the Percolation Threshold
Yuxuan Huang (Columbia University)*; Marshall Tekell (Columbia University); Hanqing Fan (Columbia University); Jingchao Qin (Columbia University); Sanat Kumar (Columbia University); Ngai Yin Yip (Columbia University)

10:00 AM (Capstone, Tuesday)
Oral 80 - Chelating DGA membranes enable ion-ion selectivity in radiopharmaceutical processing
Megan Sibley (Case Western Reserve University); Tuli Banik (Case Western Reserve University); Pavithra H. A. Kankanamalage (Brookhaven National Laboratory); Maura Sepesy (Case Western Reserve University); Alexa Ford (Case Western Reserve University); Tim Yen (Case Western Reserve University); Jasmine Hatcher-Lamarre (Brookhaven National Laboratory); Cathy Cutler (Brookhaven National Laboratory); Christine E Duval (Case Western Reserve University)*

10:30 AM (Capstone, Tuesday)
Oral 81 - Selective extraction of medium-chain fatty acid from organic waste streams using supported liquid membranes
Hiroki Fukuda (University of British Columbia); Jongho Lee (University of British Columbia)*

11:00 AM (Capstone, Tuesday)
Oral 82 - Foundational Framework for Evaluating Performance of Nanofiltration-based Solute-solute Separation
Shihong Lin (Vanderbilt University)*; Ruoyu Wang (Vanderbilt University)

11:30 AM (Capstone, Tuesday)
Oral 83 - A Novel Membrane Solvent Extraction Process Enabling Highly Efficient Separation and Recovery of Critical Materials from End-of-Life Lithium-Ion Batteries
Shailesh Singh Dangwal (Oak Ridge National Lab)*; Syed Islam (Oak Ridge National Laboratory); Priyesh Wagh (Oak Ridge National Lab); John Klaehn (Idaho National Laboratory); Ramesh Bhave (Oak Ridge National Lab)

12:00 PM (Capstone, Tuesday)
Oral 84 - Isolation of Rare Earth Ions Using Electrodialysis or Nanofiltration
Merlin Bruening (Notre Dame)*; Dong Ding (University of Notre Dame); Jun Wang (University of Notre Dame)

Oral 15 – Emerging Materials for Liquid Separations

9:30 AM – 12:30 PM, Private Dining

Session Chairs

Oishi Sanyal
West Virginia University

Volodymyr Tarabara
Michigan State University

9:30 AM (Private Dining, Tuesday)
Oral 61 - Solvent-Free Carrier-Based Membranes for Selective Ion Separations
Jay Werber (University of Toronto)*

10:00 AM (Private Dining, Tuesday)
Oral 62 - Structure and transport studies of carbon molecular sieve membranes for wastewater treatment
Young Hee Yoon (Georgia Institute of Technology)*; Ryan Lively (Georgia Tech)

10:30 AM (Private Dining, Tuesday)
Oral 63 - Differences in water and vapor transport through angstrom-scale pores in atomically thin membranes
Piran Kidambi (Vanderbilt University)*

11:00 AM (Private Dining, Tuesday)
Oral 64 - Removal of Micro- and Nano-plastics from Water using Liquid-Liquid Extraction
Ashish Srivastava (The University of Alabama)*; Milad Rabbani Esfahani (The University of Alabama);

Jason Bara (University of Alabama)

11:30 AM (Private Dining, Tuesday)
Oral 65 - Synthesis and fabrication of mesoporous membranes with enhanced mechanical properties from amphiphilic multiblock polymers via self-assembly and nonsolvent-induced phase separation
Ben Pedretti (University of Texas at Austin)*

12:00 PM (Private Dining, Tuesday)
Oral 66 - NONE

Oral Sessions - Tuesday

Presenting authors are indicated by an asterisk*.

Oral 16 – Molecular and Process Modeling

2:00 PM – 5:00 PM, Druid

Session Chairs

Shihong Lin
Vanderbilt University

Akshay Deshmukh
Massachusetts Institute of Technology

2:00 PM (Druid, Tuesday)
Oral 91 - Molecular Dynamics Simulations of Charge Interactions in Polyamide Membranes
Suwei Liu (Northwestern University); Sinan Keten (Northwestern University); Richard Lueptow (Northwestern University)*

2:30 PM (Druid, Tuesday)
Oral 92 - Water Model Choice for Molecular Dynamics Simulations of Polyamide Membranes
Suwei Liu (Northwestern University)*; Sinan Keten (Northwestern University); Richard Lueptow (Northwestern University)

3:00 PM (Druid, Tuesday)
Oral 93 - Physical mechanism of water transport in reverse osmosis membranes: Solution-diffusion or pore-flow?
Li Wang (Tongji University)*; Mohammad Heiranian (Yale University); Menachem Elimelech (Yale University)

3:30 PM (Druid, Tuesday)
Oral 94 - Machine Learning Modeling of Intermittent Reverse Osmosis Desalination System Operation via Transfer Learning and Long Short-term Memory
Yang Zhou (Shanghai University); Nora Marki (University of California, Los Angeles); Christian Aguilar (University of California, Los Angeles); Kwan Lam Yip (University of California, Los Angeles); Yakubu Jarma (University of California, Los Angeles); Bilal Khan (California State University, San Bernardino); Yoram Cohen (University of California, Los Angeles)*

4:00 PM (Druid, Tuesday)
Oral 95 - Impact of dynamical correlations on salt transport in solvated ion exchange membranes
Nico Marioni (The University of Texas at Austin)*; Akhila Rajesh (The University of Texas at Austin); Zidan Zhang (The University of Texas at Austin); Harnoor Sachar (The University of Texas at Austin); Benny Freeman (The University of Texas at Austin); Venkat Ganesan (The University of Texas at Austin)

4:30 PM (Druid, Tuesday)
Oral 96 - Monovalent Ion Selectivity of Cellulose Acetate Membranes
Paul R Irving (University of Texas at Austin)*; Kevin Reimund (University of Texas at Austin); Everett S Zofchak (The University of Texas at Austin); Nico Marioni (The University of Texas at Austin); Benny Freeman (The University of Texas at Austin); Venkat Ganesan (The University of Texas at Austin)

Oral 17 – Organic Solvent Separations

2:00 PM – 5:00 PM, Yellow Hammer

Session Chairs

Michele Galizia
Oklahoma University

Neel Rangnekar
Exxonmobil

2:00 PM (Yellow Hammer, Tuesday)
Oral 97 - Transport Property Modulation via Non-Aqueous Solvent Specific Behavior in Cross-Linked Membranes
Charles Leroux (University of Virginia)*; Patrick McCormack (University of Virginia); Geoffrey M Geise (University of Virginia); Gary Koenig (University of Virginia)

2:30 PM (Yellow Hammer, Tuesday)
Oral 98 - Engineering organic solvent reverse osmosis performance of vapor phase infiltrated organic-inorganic hybrid membranes
Yi Ren (Georgia Institute of Technology)*

3:00 PM (Yellow Hammer, Tuesday)
Oral 99 - Development of a high-performance 30nm NIPS PTFE membrane
Karl Ashkar (Pall Corporation)*; Siavash Darvishmanesh (Pall Corporation); Steven Gardner (Pall Corporation)

3:30 PM (Yellow Hammer, Tuesday)
Oral 100 - Crosslinked thin film composite membranes for industrial organic complex mixtures separations
Yacine Feliachi (Georgia Tech)*; Ryan Lively (Georgia Tech); M.G. Finn (Georgia Institute of Technology); Ankana Roy (Georgia Tech)

4:00 PM (Yellow Hammer, Tuesday)
Oral 101 - Ultrathin polytriazole membranes for energy-intensive separation
Stefan Chisca (KAUST)*; Suzana Nunes (King Abdullah University of Science and Technology (KAUST))

4:30 PM (Yellow Hammer, Tuesday)
Oral 102 - Unveiling the role of sorption, diffusion, interactions and plasticization in organic solvent transport in dense OSN/OSRO membranes
Michele Galizia (University of Oklahoma)*

Oral 18 – Membrane Characterization

2:00 PM – 5:00 PM, Black Warrior

Session Chairs
Abhishek Mondal
Donaldson Inc.

Sudhir Ravula
The University of Alabama

2:00 PM (Black Warrior, Tuesday)
Oral 103 - Improved Mechanical Compliance in Bicontinuous Cubic Lyotropic Membranes Based on a Cross-Linking Gemini Monomer via Copolymerization with a Non-Cross-Linkable Analog
Lauren N Bodkin (University of Colorado Boulder); Patrick Li (University of Colorado Boulder); Samantha Dyer (University of Colorado Boulder); Zachary Krajnak (University of Colorado Boulder); John Malecha (University of Colorado Boulder); Richard D. Noble (University of Colorado, Boulder)*; Douglas Gin (Department of Chemistry, University of Colorado, Boulder)

2:30 PM (Black Warrior, Tuesday)
Oral 104 - Advanced metrology, molecular dynamics, and machine learning for determining ion activity coefficients in ion-exchange membranes
Christopher G Arges (The Pennsylvania State University)*; Hishara Keshani Gallange Dona (Louisiana State University); Teslim Olayiwola (Louisiana State University); Luis Briceno-Mena (Louisiana State University); Mario Ramos-Garces (The Pennsylvania State University); Revati Kumar (Louisiana State University); Jose Romagnoli (Louisiana State University)

3:00 PM (Black Warrior, Tuesday)
Oral 105 - Mathematical Modeling of CO_2 Facilitated Transport across Amine-functionalized Polymeric Membranes with Direct Operando Observation of Amine Carrier Saturation
Casey O'Brien (University of Notre Dame)*

3:30 PM (Black Warrior, Tuesday)
Oral 106 - Unveiling the mechanism of light gas transport through configurational free volume in glassy polymers
William Box (University of Oklahoma)*; Zihan Huang (University of Notre Dame); Ruilan Guo (University of Notre Dame); Michele Galizia (University of Oklahoma)

4:00 PM (Black Warrior, Tuesday)
Oral 107 - Characterizing dielectric properties and water molecule dynamics using dielectric relaxation spectroscopy
Sean M Bannon (University of Virginia)*; Charles Leroux (University of Virginia); Beatrice M Tremblay (University of Virginia); Geoffrey M Geise (University of Virginia)

4:30 PM (Black Warrior, Tuesday)
Oral 108 - A Comprehensive OMICs Platform for Detecting Biofouling in Full-Scale Reverse Osmosis Facilitates
Weiliang Bai (University of Texas at Austin)*; Manish Kumar (University of Texas at Austin)

Oral 19 – Electrochemical Separations 2

2:00 PM – 5:00 PM, Capstone

Session Chairs

Hee Jeung Oh
Penn State University

Shudipto Dishari
University of Nebraska-Lincoln

Piran Kidambi
Vanderbilt University

William Tarpeh
Stanford University

2:00 PM (Capstone, Tuesday)
Oral 109 - A new polymer platform for anion exchange membranes based on Friedel-Crafts polymerization
Lingyu Yang (University of Notre Dame); Chengkai Fan (University of Notre Dame); Haifeng Gao (University of Notre Dame); Jennifer L Schaefer (University of Notre Dame)*

2:30 PM (Capstone, Tuesday)
Oral 110 - Highly Charged Ion-Exchange Membranes for Treating Concentrated Salt Solutions via Electrodialysis
Jovan Kamcev (University of Michigan)*

3:00 PM (Capstone, Tuesday)
Oral 111 - Production of polyphenols-enriched cranberry juice through electrodialysis with filtration membrane: How physicochemical properties of filtration membranes impact anthocyanins migration and selectivity during electrodialysis
Eva Revellat (Université Laval; Institute of Nutrition and Functional Foods)*; Laurent Bazinet (Laval University, Institute on Nutrition and Functional Food (INAF))

3:30 PM (Capstone, Tuesday)
Oral 112 - Correlation between volatile fatty acid transport properties in anion exchange membranes and volatile fatty acid transport during electrodialysis
Holly M Haflich (University of North Carolina- Chapel Hill)*; Josh Singleton (University of North Carolina-Chapel Hill); Orlando Coronell (University of North Carolina at Chapel Hill)

4:00 PM (Capstone, Tuesday)
Oral 54 - Bipolar membrane electrodialysis for direct LiOH production from concentrated brines
Aubrey Quigley (The University of Texas at Austin)*; Kevin Reimund (University of Texas at Austin); Nathaniel Lynd (The University of Texas at Austin); Benny Freeman (The University of Texas at Austin)

4:30 PM (Capstone, Tuesday)
Oral 114 - Revolutionizing Ion Transport: Dynamic Channel-Size Tuning in MXene Transistor Membranes for Selective Separation and Enhanced Control
Aaditya A Pendse (Lawrence Livermore National Labortory)*; Aleksandr Noy (Lawrence Livermore National Laboratory)

Oral 20 – Membrane Distillation and Pervaporation

2:00 PM – 5:00 PM, Private Dining

Session Chairs

Lucy Mar Camacho
Texas A&M University-Kingsville

Tiezheng Tong
Colorado State University

2:00 PM (Private Dining, Tuesday)
Oral 115 - Anti-scalants for mitigating silica scaling in membrane desalination: Effects of molecular structure and membrane process
Yiqun Yao (Colorado State University)*; Tiezheng Tong (Colorado State University)

2:30 PM (Private Dining, Tuesday)
Oral 116 - New approaches to mitigate wetting and scaling in membrane distillation
Shihong Lin (Vanderbilt University)*

3:00 PM (Private Dining, Tuesday)
Oral 117 - Hydrophobic Hollow Fiber-Membrane for Spacesuit Cooling
Dibakar Bhattacharyya (University of Kentucky)*; M.Arif Khan (University of Kentucky); Glenn Lipscomb (University of Toledo, Toledo); Andrew Lin (University of Kentucky); Morgan Abney (NASA Langley Research Center)

3:30 PM (Private Dining, Tuesday)
Oral 118 - Radiofrequency responsive

membranes for membrane distillation applications

Md Hasib Al Mahbub (Texas Tech University); Mohammad Saed (Texas Tech University); Mahdi Malmali (Texas Tech University)*

4:00 PM (Private Dining, Tuesday)

Oral 119 - Materials and design to enhance transport in membrane distillation

Waqas Alum (Purdue University)*; David M Warsinger (Purdue)

4:30 PM (Private Dining, Tuesday)

Oral 120 - Omniphobic electrospun membranes for treatment of hydraulic fracturing produced water through membrane distillation

Ranil Wickramasinghe (University of Arkansas)*; YuHsuan Chiao (University of Arkansas)

Oral Sessions - Wednesday

Presenting authors are indicated by an asterisk*.

Oral 21 – Awards Session

9:30 AM – 12:30 PM, Druid

Session Chairs
Evan Hatakeyama
Chevron

Ryan Lively
Georgia Tech

9:30 AM (Druid, Wednesday)
Oral 121 - Challenges and opportunities in ultra-high pressure reverse osmosis operation
Jishan Wu (UCLA)*; Eric Hoek (UCLA)

10:00 AM (Druid, Wednesday)
Oral 122 - Kinetic Control of Angstrom-scale Porosity in Graphene for Direct Scalable Synthesis of Atomically Thin Proton Exchange Membranes
Nicole K Moehring (Vanderbilt University)*

10:30 AM (Druid, Wednesday)
Oral 123 - Data-driven predictions of complex mixture permeation in polymer membranes
Youngjoo Lee (Georgia Tech)*; Lihua Chen (Georgia Tech); Rampi Ramprasad (); Ryan Lively (Georgia Tech)

11:00 AM (Druid, Wednesday)
Oral 124 - Modification and Synthesis of Polyamide Reverse Osmosis Membranes for the Separation of Small, Neutral Molecules
Shahriar Habib (The University of Alabama); Madison A Wilkins (The University of Alabama); Bryn Larson (The University of Alabama); Steven T Weinman (The University of Alabama)*

11:30 AM (Druid, Wednesday)
Oral 125 - High-Performance Non-Aging Rubbery Thin Film Composite Membranes for Post-Combustion CO_2 Capture: From Material Synthesis to Field Demonstration
Lingxiang Zhu (National Energy Technology Laboratory)*; Victor A Kusuma (National Energy Technology Laboratory); David Hopkinson (National Energy Technology Laboratory)

12:00 PM (Druid, Wednesday)
Oral 126 - Tuning 2D materials channels for sub-nano scale separation: from interlayers to in-plane pores
Jie Shen (King Abdullah University of Science and Technology)*

Oral 22 – Membranes for Gas Separations

9:30 AM – 12:30 PM, Yellow Hammer

Session Chairs
Ali Rownaghi
Cleveland State University

Surya Tiwari
National Energy Technology Laboratory

9:30 AM (Yellow Hammer, Wednesday)
Oral 145 - Co^{2+}-based supramolecular networks achieving extraordinary CO_2/O_2 selectivity
Taliehsadat Alebrahim (The State University of New York at Buffalo)*; Narjes Esmaeili (The State University of New York at Buffalo); Gengyi Zhang (The State University of New York at Buffalo); Thien N Tran (University at Buffalo, The State University of New York); Haiqing Lin (The State University of New York at Buffalo)

10:00 AM (Yellow Hammer, Wednesday)
Oral 146 - Crown Ether-Based Polyimides for Sour Gas Separation
Dana Wong (Aramco Americas)*; Elizabeth Haddad (Aramco Americas); John Yang (Aramco Americas); Patrick Wright (Aramco Americas)

10:30 AM (Yellow Hammer, Wednesday)
Oral 147 - Modeling Gas Permeation in Amorphous Polymer Membranes Using Non-Equilibrium Molecular Dynamics Simulations
Mohammed N Al Otmi (University of Florida)*; Janani Sampath (University of Florida)

11:00 AM (Yellow Hammer, Wednesday)
Oral 148 - Selective H_2S/CO_2 Separation using Sterically Hindered Amine Membranes
Shraavya Rao (Ohio State University)*; Xuepeng Deng (The Ohio State University); Yang Han (The Ohio State University); Li-Chiang Lin (The Ohio State University); Winston Ho (Ohio State University)

11:30 AM (Yellow Hammer, Wednesday)
Oral 149 - Sweetening Sour Gas with Hybrid Membranes
Patrick T Wright (Aramco Americas)*; John Yang

(Aramco Americas)

12:00 PM (Yellow Hammer, Wednesday)
Oral 150 - Identifying key factors to unlock plasticization resistance in microporous polymer membranes
Katherine Mizrahi Rodriguez (MIT); Sharon Lin (MIT); Albert Wu (MIT); Kayla Storme (MIT); Taigyu Joo (Massachusetts Institute of Technology); Aristotle F Grosz (MIT)*; Naksha Roy (MIT); Duha Syar (MIT); Francesco Benedetti (MIT); Zachary Smith (MIT)

Oral 23 – Contaminant Removal from Water

9:30 AM – 12:30 PM, Black Warrior

Session Chairs
Jung Min Kim
University of Virginia

Boya Xiong
University of Minnesota

9:30 AM (Black Warrior, Wednesday)
Oral 133 - Polyamide Reverse Osmosis Membrane Modification with Diamines to Improve Small Neutral Molecule Rejection
Shahriar Habib (The University of Alabama)*; Madison A Wilkins (The University of Alabama); Steven T Weinman (The University of Alabama)

10:00 AM (Black Warrior, Wednesday)
Oral 134 - Electrocatalytic laser-induced graphene membrane for micropollutant removal and biofouling control
Swatantra Pratap Singh (IIT Bombay)*; Najmul Barbhuiya (IIT Bombay); Kritika Jasrapuria (IIT Bombay)

10:30 AM (Black Warrior, Wednesday)Oral 135 - Simultaneous Uranium Extraction and Brine Desalination via Adsorptive Pervaporation by A Bifunctional Graphene Oxide Membrane
Guangcheng Wang (National University of Singapore)*; Sui Zhang (National University of Singapore)

11:00 AM (Black Warrior, Wednesday)
Oral 136 - Simple and low-cost electroactive membranes for ammonia recovery
Xinyi Wang (University of California, Los Angeles)*; Sungju Im (University of California, Los Angeles)

11:30 AM (Black Warrior, Wednesday)
Oral 137 MIL-100 (Fe) Functionalized Thin-film Nanocomposite Membranes for Enhanced Removal of Uncharged Urea from Water
Tin Le (The University of Alabama)*; Milad Rabbani Esfahani (The University of Alabama)

12:00 PM (Black Warrior, Wednesday)
Oral 138 - Microcystin-LR destruction and removal from water via enzymatic biodegradation and ultrafiltration
Abelline K Fionah (University of Kentucky)*; Isabel Escobar (University of Kentucky)

Oral 24 – Membrane Fouling

9:30 AM – 12:30 PM, Capstone

Session Chairs
Daniel Miller
Twelve

Audie Thompson
U.S. Army ERDC

Ngoc Bui
University of Oklahoma

9:30 AM (Capstone, Wednesday)
Oral 139 - Energy reduction in ceramic microfiltration using supersaturated carbon-dioxide-enhanced backwash
Weiming Qi (Clemson University)*; David Ladner (Clemson)

10:00 AM (Capstone, Wednesday)
Oral 140 - The role of surfactants in membrane fouling
Naama Segev-Mark (Technion-Israel Institute of Technology)*; Kirti Sankhala (Technion-Israel Institute of Technology); David Jassby (UCLA); Eric Hoek (UCLA); Guy Z Ramon (Technion - Israel Institute of Technology)

10:30 AM (Capstone, Wednesday)
Oral 141 - In-situ measurements to differentiate homogeneous and heterogeneous nucleation in membrane distillation: effects of spacer geometry and antiscalants
Mahdi Malmali (Texas Tech University)*; Hadis Matinpour (University of California - Santa Barbara); I-Min Hsieh (Texas Tech University); Bosong Lin (Texas Tech University)

11:00 AM (Capstone, Wednesday)
Oral 142 - Antimicrobial Properties of Chitosan-Graphene Oxide Composites
Christopher S Griggs (U.S. Army ERDC)*; Audie Thompson (U.S. Army ERDC); Justin Puhnaty (U.S. Army ERDC)

11:30 AM (Capstone, Wednesday)
Oral 143 - Spacers for Mass Transfer Enhancement in Flat Sheet Membrane Modules – An Upper Bound?
Arsalan Sepehri (University of Toledo); Glenn Lipscomb (University of Toledo, Toledo)*

12:00 PM (Capstone, Wednesday)
Oral 144 - Real-Time Induced Magnetic Vibrational Based Antifouling Mechanism for Ultrafiltration (UF) membrane
Jasneet K Pala (The University of Alabama)*; Ryan Tracy (The University of Alabama); Milad Rabbani Esfahani (The University of Alabama); Nima Mahmoodi (The University of Alabama)

Oral 25 – Process Intensification and Integration

9:30 AM – 12:30 PM, Private Dining

Session Chairs

Marie-Eve Langevin
Ameridia Inc.

Mahdi Malmali
Texas Tech University

Albert Kim
University of Hawai'i

Jai Wei Chew
Nanyang Technological University

9:30 AM (Private Dining, Wednesday)
Oral 127 - Thermally and Chemically Stable Metallic Membrane Reactors for H_2 Generation with CO_2 Capture
Hani A.E. Hawa (Precision Combustion, Inc.)*; Kyle Hawley (Precision Combustion, Inc.); Damian Agi (University of Notre Dame); Alexander Dowling (University of Notre Dame)

10:00 AM (Private Dining, Wednesday)
Oral 128 - Specific Energy Analysis of Using Fertilizer-Based Liquid Desiccants to Dehumidify Indoor Plant Environments
Sarah Moussaddy (Oakland University)*; Jonathan Maisonneuve (Oakland University); Sandeep Aryal (Oakland University)

10:30 AM (Private Dining, Wednesday)
Oral 129 - Feasibility of Membrane-adsorption Hybrid Process for Direct Air Capture
Yi-Chen Huang (The Ohio State University)*; Yang Han (The Ohio State University); Winston Ho (The Ohio State University)

11:00 AM (Private Dining, Wednesday)
Oral 130 - Electrospray 3D Printing to Unlock Customized Membrane Performance
Mert Can Hacifazlioglu (University of Connecticut)*; Jeffrey R McCutcheon (University of Connecticut)

11:30 AM (Private Dining, Wednesday)
Oral 131 - Intensified membrane distillation for treating oilfield produced waters from unconventional wells: Energetics analysis and pilot demonstration
Mahdi Malmali (Texas Tech University)*; Md Hasib Al Mahbub (Texas Tech University); I-Min Hsieh (Texas Tech University); Bosong Lin (Texas Tech University)

12:00 PM (Private Dining, Wednesday)
Oral 132 - Low concentration carbon dioxide capture using MEEP-based membranes
Birendra Adhikari (Idaho National Laboratory)*; Amit Nilkar (Idaho National Laboratory); Christopher Orme (Idaho National Laboratory); John Klaehn (Idaho National Laboratory); Hyoenseok Lee (Idaho National Laboratory)

Oral Sessions - Wednesday

Presenting authors are indicated by an asterisk*.

Oral 26 – High Salinity Streams and Electrochemical Separations 3

2:00 PM – 5:00 PM, Druid

Session Chairs

Kofi Christie
Louisiana State University

Bianca Souza Chaves
University of Arizona

Jonathan Brant
University of Wyoming

2:00 PM (Druid, Wednesday)
Oral 151 - Sulfonated Polystyrene Membranes Near the Percolation Threshold for Hypersaline Electrodialysis Desalination
Yuxuan Huang (Columbia University)*; Marshall Tekell (Columbia University); Hongxu Chen (Columbia University); Sanat Kumar (Columbia University); Ngai Yin Yip (Columbia University)

2:30 PM (Druid, Wednesday)
Oral 152 - Controlled Localized Metal Organic Framework Synthesis on Anion Exchange Membranes
Harm Wiegerinck (University of Twente)*

3:00 PM (Druid, Wednesday)
Oral 153 - Assessing the Suitability of Nanofiltration for Pretreatment of Lithium-Enriched Brines
Juan Zhai (Texas Tech University)*; Yuexiao Shen (Texas Tech University); Mahdi Malmali (Texas Tech University); rajesh khare (Texas Tech University); Amrika Deonarine (Texas Tech University); Shubhra Bhattacharjee (Texas Tech University); Adegbola Balogun (Texas Tech University)

3:30 PM (Druid, Wednesday)
Oral 154 - Ion transport in ion exchange membranes: water confinement effects at low hydration
José Carlos Díaz (University of Michigan)*

4:00 PM (Druid, Wednesday)
Oral 155 - Extending the reverse osmosis to osmotic mediation for brine management
Ali Naderi Beni (Purdue University)*; Iraj Ghofrani (Sharif University of Technology); Ali Mousavi (Sharif University of Technology); David M Warsinger (Purdue)

4:30 PM (Druid, Wednesday)
Oral 156 - Understanding mixed salt partitioning in model poly(ethylene oxide)-based membrane materials
Everett S Zofchak (The University of Texas at Austin)*; Aubrey Quigley (The University of Texas at Austin); Jordyn Yoh (The University of Texas at Austin); Venkat Ganesan (The University of Texas at Austin); Benny Freeman (The University of Texas at Austin)

Oral 27 – Water Reuse

2:00 PM – 5:00 PM, Yellow Hammer

Session Chairs

Sage Hiibel
University of Nevada-Reno

Vicky Karanikola
University of Arizona

2:00 PM (Yellow Hammer, Wednesday)
Oral 157 - Advancements in Water Treatment: Concentrating RO Residuals Contaminated with PFAS through Closed Circuit RO and Membrane Distillation
Tae Lee (US EPA)*; Mallikarjuna Nadagouda (Wright State University); Hyunsik Kim (University of Cincinnati)

2:30 PM (Yellow Hammer, Wednesday)
Oral 158 - Backwashable hollow-fibre membranes enable ultrahigh recovery nanofiltration process for effluent reuse
Chidiebere S Nnebuo (Ben-Gurion University of the Negev)*; Oded Nir (Ben-Gurion University of the Negev)

3:00 PM (Yellow Hammer, Wednesday)
Oral 159 - A membrane-based multi-barrier approach to potable water reuse and waste heat recovery
Andrea Achilli (University of Arizona)*; Zachary Binger (The University of Arizona); Mukta Hardikar (The University of Arizona); Sage Hiibel (University of Nevada, Reno); Eric Marchand (University of Nevada, Reno); Amy Childress (University of Southern California)

3:30 PM (Yellow Hammer, Wednesday)
Oral 160 - Transport mechanisms of fouled engineering-scale water reuse reverse

osmosis membranes
Bianca M. Souza Chaves (The University of Arizona)*; Mohammed Alhussaini (The university of Arizona); Varinia V Felix (The University of Arizona); Andrea Achilli (University of Arizona)

4:00 PM (Yellow Hammer, Wednesday)
Oral 161 - Selective removal of selenium from saline water using different MOFs and MOF-incorporated NF membrane
Sweta A Modak (The University of Alabama)*; Milad Rabbani Esfahani (The University of Alabama)

4:30 PM (Yellow Hammer, Wednesday)
Oral 162 - NONE

Oral 28 – Bioinspired and Biomimetic Membranes

2:00 PM – 5:00 PM, Black Warrior

Session Chairs
Cassandra Porter
Auburn University

Yuexiao Shen
Texas Tech University

Harekrushna Behera
University of Texas at Austin

2:00 PM (Black Warrior, Wednesday)
Oral 163 - Tunable membranes incorporating artificial water channels for high-performance water reverse osmosis desalination
Mihail Barboiu (Institut Europeen des Membranes)*

2:30 PM (Black Warrior, Wednesday)
Oral 164 - Development of Nanofiltration membranes incorporating Artificial Water Channel
Omar Samhari (Université de Montpellier)*; Mihail Barboiu (Institut Europeen des Membranes)

3:00 PM (Black Warrior, Wednesday)
Oral 165 - Effect of Nanopatterning on E. Coli Fouling and Concentration Polarization
Lauren M Ward (The University of Alabama)*; Georgie Fickling (The University of Alabama); Urmi Roy (The University of Alabama); Steven T Weinman (The University of Alabama); Rushabh Shah (University of Massachusetts Amherst); Jessica Schiffman (University of Massachusetts Amherst)

3:30 PM (Black Warrior, Wednesday)
Oral 166 - Biomimetic Supramolecular Channels for Sustainable Extraction of Rare Earth Elements
Harekrushna Behera (The University of Texas at Austin)*; Manish Kumar (University of Texas at Austin); Hyeonji Oh (University of Texas at Austin); Laxmicharan Samineni (University of Texas at Austin); Tyler J Duncan (The University of Texas at Austin); Venkat Ganesan (The University of Texas at Austin)

4:00 PM (Black Warrior, Wednesday)
Oral 167 - Biomimetic membranes with high water vapor permeance and protective capacity for next generation breathable protective fabrics
Hyeonji Oh (University of Texas at Austin)*; Yu-Ming Tu (MIT); Behzad Mehrafrooz (UIUC); Veda Sheersh Boorla (Pennsylvania State University); Harekrushna Behera (University of Texas at Austin); Costas Maranas (Pennsylvania State University); Aleksei Aksimentiev (University of Illinois at Urbana-Champaign); Manish Kumar (University of Texas at Austin)

4:30 PM (Black Warrior, Wednesday)
Oral 168 - NONE

Oral 29 – Membrane and Module Optimization

2:00 PM – 5:00 PM, Capstone

Session Chairs
Hani Hawa
Precision Combustion, Inc.

2:00 PM (Capstone, Wednesday)
Oral 169 - Random polyampholyte amphiphilic copolymers for membranes with tunable selectivity and excellent fouling resistance
Ayse Asatekin (Tufts University)*; Luca Mazzaferro (Tufts University)

2:30 PM (Capstone, Wednesday)
Oral 170 - Co-transport of methanol and formate in crosslinked phenyl acrylate-based ion exchange membranes: Effect of steric hindrance
Jung Min Kim (University of Virginia)*; Yihung Lin (Auburn University); Sean M Bannon (University of Virginia); Geoffrey M Geise (University of Virginia); Bryan S Beckingham (Auburn University)

3:00 PM (Capstone, Wednesday)

Oral 171 - Acoustic actuated 3D printed bubble-trapping structures for in-situ mitigation of membrane fouling

Kieran R Fung (Membrane Science, Engineering, and Technology (MAST) Center, Department of Mechanical Engineering, University of Colorado Boulder)*; Shouhong Fan (University of Colorado at Boulder); Jason P. Killgore (Applied Chemicals and Materials Division, National Institute of Standards and Technology, Boulder, CO); Yifu Ding (Membrane Science, Engineering, and Technology (MAST) Center, Department of Mechanical Engineering, University of Colorado Boulder); Xiaoyun Ding (Department of Mechanical Engineering, University of Colorado Boulder)

3:30 PM (Capstone, Wednesday)

Oral 172 - High-performance large-area polymeric membranes with chlorine tolerance for water purification by pressure-driven distillation

Duong T. Nguyen (University of Colorado Boulder)*; Kian P Lopez (University of Colorado Boulder); Shouhong Fan (University of Colorado Boulder); Yifu Ding (Membrane Science, Engineering, and Technology (MAST) Center, Department of Mechanical Engineering, University of Colorado Boulder); Jongho Lee (University of British Columbia); Anthony Straub (University of Colorado Boulder)

4:00 PM (Capstone, Wednesday)

Oral 173 - Graphene Oxide and Sulfonated graphene oxide embedded Polyamide-polysulfone based Thin-film Nanocomposite Sea water Reverse Osmosis membranes

Asim K Ghosh (Bhabha Atomic research centre, Trombay, Mumbai)*; Rutuja Bhoje (Institute of Chemical Technology); Parag Nemade (Institute of chemical Technology)

4:30 PM (Capstone, Wednesday)

Oral 174 - Materials and design to enhance transport in membrane distillation

Waqas Alam (Purdue University*; David M Warsinger (Purdue); Hamid Juybari (Purdue University)

Poster Sessions - Monday

Presenting authors are indicated by an asterisk*.

Poster Session (Ballroom)

7:00 PM – 10:00 PM

Session Chairs

Yuexiao Shen
Texas Tech University

Yu-Ming Tu
Massachusetts Institute of Technology

Siamak Nejati
University Of Nebraska-Lincoln

Poster 1 - Synergistic Effect of Combining UIO-66 and Mxene nanosheets in Pebax Mixed Matrix Membranes for CO_2 Capture and Separation
Eyasu Ajebe (National Taiwan University of Science and Technology)*

Poster 2 - Polymers of intrinsic microporosity containing aryl-phthalimide moieties: synthesis, modeling, and membrane gas transport properties
Alain Tundidor (Pontificia Universidad Catolica de Chile)*; Fidel Rodriguez (Pontificia Universidad Catolica de Chile); Claudio Terraza (Pontificia Universidad Catolica de Chile); Bibiana Comesaña (Universidad de Valladolid); Ana Lilian Montero (Universidad Tecnologica Metropolitana)

Poster 3 - New polyimides of intrinsic microporosity: synthesis, characterization and study of properties
María Victoria Velázquez (Pontificia Universidad Católica de Chile)*; Alain Tundidor (Pontificia Universidad Catolica de Chile); Claudio Terraza (Pontificia Univesidad Católica de Chile); Fidel Rodríguez (Pontificia Universidad Católica de Chile)

Poster 4 - Ultraselective Carbon Molecular Sieve Membranes Derived from Solution-Processable Aramids
Gaurav Iyer (University of Maryland); Chen Zhang (University of Maryland)*

Poster 5 - Specific Ion Selectivity in Sulfonated Polystyrene Membranes Near the Percolation Threshold
Yuxuan Huang (Columbia University)*; Marshall Tekell (Columbia University); Hanqing Fan (Columbia University); Jingchao Qin (Columbia University); Sanat Kumar (Columbia University); Ngai Yin Yip (Columbia University)

Poster 6 - Metal ion transport in dense membranes containing zwitterionic sulfobetaine methacrylate and hydrophobic phenyl acrylate
Yihung Lin (Auburn University)*; Jung Min Kim (University of Virginia); Bryan S Beckingham (Auburn University)

Poster 7 - Flexible operation of RO desalting in the mode of partial concentrate recycle
Yakubu Jarma (University of California, Los Angeles); Yoram Cohen (Professor)*; Beau Akhavain (University of California, Los Angeles)

Poster 8 - The composite water swollen spiral wound membrane module for raw biogas purification
Petra Wojnarova (Institute of Chemical Process Fundamentals of the Czech Academy of Sciences); Jiri Rusin (Institute of Chemical Process Fundamentals of the Czech Academy of Sciences); Panagiotis Basinas (Institute of Chemical Process Fundamentals of the Czech Academy of Sciences); Martin Kostejn (Czech Academy of Sciences); Petr Stanovsky (Czech Academy of Sciences); Pavel Izak (Institute of Chemical Process Fundamentals of the Czech Academy of Sciences); Albert Kim (U. of Hawai'i)*

Poster 9 - Oriented Metal-Organic Framework Nanosheets in Mixed-matrix Membranes for Simultaneous H_2S and CO_2 Removal from Natural Gas
Biplab Joarder (KAUST)*; Shuvo Jit Datta (KAUST); Mohamed Eddaoudi (KAUST)

Poster 10 - Highly Porous Membrane Adsorbents Containing Polyethyleneimine for Direct Air Capture
Shweta Singh (University at Buffalo)*; Haiqing Lin (The State University of New York at Buffalo); Thien N Tran (University at Buffalo, The State University of New York)

Poster 11 - Extracting Lithium from Hypersaline Salt-lakes with Selective Electrodialysis and

Nanofiltration
Zi Hao Foo (Massachusetts Institute of Technology)*; John Lienhard (Massachusetts Institute of Technology)

Poster 12 - Lithium and Sodium Sorption in Crosslinked Bisphenol A Ethoxylate Diacrylate-based Cation Exchange Membranes
Beatrice M Tremblay (University of Virginia)*; Geoffrey M Geise (University of Virginia); Sean M Bannon (University of Virginia); Jung Min Kim (University of Virginia)

Poster 13 - Analysis of Manufacturing Methods of Ion Exchange Membranes for Desalination
Anna Harris (University of Virginia)*; Geoffrey M Geise (University of Virginia); Inara Oliveira do Carmo Nascimento (University of Virginia); Jeffrey R McCutcheon (University of Connecticut); Deniz Ipekci (University of Connecticut)

Poster 14 - A Click Chemistry-Based Membrane Platform for Selective Separations
Lianna Johnson (CWRU)*; Bernadette Schneider (CWRU); Christine E Duval (Case Western Reserve University)

Poster 15 - Upstream radiopharmaceutical purification: purifying metals in strong acids
Maura Sepesy (Case Western Reserve University)*; Amy Che (Case Western Reserve University); Christine E Duval (Case Western Reserve University)

Poster 16 - Functional Hollow Fibers Made from Polymers of Intrinsic Microporosity for Membrane-Based Separations
Philippe Jean-Baptiste (Massachusetts Institute of Technology)*; Zachary Smith (MIT)

Poster 17 - Micro-phase Separated Blends of poly(PFMMD-co-CTFE) and Teflon AF2400 exhibiting tunable helium permeability/selectivity
Lucas Condes (University of Oklahoma)*; Michele Galizia (University of Oklahoma); Shivam Patel (University of Oklahoma); Bryce Goodin (University of Oklahoma)

Poster 18 - Evaluating the Experimental Uncertainty in Gas and Vapor Sorption/Adsorption Measurements: Fundamental Considerations and Experimental Design Implications
William Box (University of Oklahoma)*; Matthew T Webb (University of Oklahoma); Michele Galizia (University of Oklahoma)

Poster 19 - Characterizing dielectric permittivity properties and water molecule dynamics using dielectric relaxation spectroscopy
Sean M Bannon (University of Virginia)*; Charles Leroux (University of Virginia); Beatrice M Tremblay (University of Virginia); Geoffrey M Geise (University of Virginia)

Poster 20 - Extending the effects of crosslinks during free volume manipulation of amine-functionalized PIM-1 membranes for gas separation
Taigyu Joo (Massachusetts Institute of Technology)*; Sungsool Wi (National High Magnetic Field Laboratory); Zachary Smith (MIT)

Poster 21 - Advancing Nanoparticle Biotherapeutic Purification for Viral Vector Bioprocessing Applications Utilizing Microfiltration Membranes
Mara Leach (University of Kentucky)*; Carter Pryor (Univerisity of Kentucky); Gosia Chwatko (University of Kentucky); Dibakar Bhattacharyya (U. Kentucky)

Poster 22 - Designing membranes using bottlebrush poly(1,3-dioxolane)acetate for CO_2 capture
Thien N Tran (University at Buffalo, The State University of New York)*; Fathy Attia (University at Buffalo, The State University of New York); Haiqing Lin (The State University of New York at Buffalo)

Poster 23 - Deep Eutectic Solvents Incorporated Hollow Fiber Membrane Contactor Enabling Energy-efficient CO_2 Capture
Syed Islam (Oak Ridge National Laboratory)*; Zachary Coin (Oak Ridge National Laboratory); Thomas Knight (Vanderbilt University); Md Arifuzzaman (Oak Ridge National Laboratory); Gernot Rother (Oak Ridge National Laboratory); Vera Bocharova (Oak Ridge National Laboratory); Robert Sacci (Oak Ridge National Laboratory); Jacek Jakowski (Oak Ridge National Laboratory); Jingsong Huang (Oak Ridge National Laboratory); Ilia Ivanov (Oak Ridge National Laboratory); Ramesh Bhave (Ornl); Tomonori Saito (Oak Ridge National Laboratory); David Sholl (Oak Ridge National Laboratory)

Poster 24 - Purification of circular RNA by ultrafiltration
Karen d Guillen Cuevas (Clemson University); Marc Birtwistle (Clemson University); Scott Husson (Clemson University)*

Poster 25 - Hollow fiber membrane supported metal organic framework-based nanopacked bed for gas/vapor separation
Yufeng Song (New Jersey Institute of Technology); Kamalesh K Sirkar (New Jersey Institute of Technology)*

Poster 26 - Application of Iron-Functionalized Membranes Towards Oxidative Degradation of Chloro-Organic Pollutants
Amirmohammad Sakhaei (University of Kentucky); Rollie Mills (U. Kentucky); Thomas Dziubla (University of Kentucky); Dibakar Bhattacharyya (U. Kentucky)*

Poster 27 - High-performance Gas Separation Membranes derived from Polyimides having Polyethylene-oxide Moieties
Matthew T Webb (University of Oklahoma)*; Laura Matensanz-Niño (University of Valladolid); Alfonso González-Ortega (University of Valladolid); Laura Palacio (University of Palacio); Cristina Álvarez (University of Valladolid); Michele Galizia (University of Oklahoma)

Poster 28 - Development of Crosslinked Polyethylene Glycol Membranes Using Click Chemistry for Sour Gas Separations
Elizabeth Haddad (Aramco Americas)*; Dana Wong (Aramco Americas); John Yang (Aramco Americas); Patrick Wright (Aramco Americas)

Poster 29 - Electrodialysis for Nitrogen Based Nutrient Recovery
Hyuck Joo Choi (Georgia Institute of Technology)*; Lily Hawk (Georgia Institute of Technology); Paul A Kohl (Georgia Institute of Technology); Marta Hatzell (Georgia Institute of Technology)

Poster 30 - Effect of Nanopatterning on E. Coli Fouling and Concentration Polarization
Lauren M Ward (The University of Alabama)*; Georgie Fickling (The University of Alabama); Urmi Roy (The University of Alabama); Steven T Weinman (The University of Alabama); Rushabh Shah (University of Massachusetts Amherst); Jessica Schiffman (University of Massachusetts Amherst)

Poster 31 - Water vapor transport in carbon molecular sieve membranes and its gas separation performance in the presence of humidity
Horacio Lopez Marques (The University of Texas at Austin)*

Poster 32 - Multicomponent transport of methanol and carboxylates in PEGDA-SPMAK/PEGPEA cation exchange membranes
Pravin Parasakthi Aravindhan (Auburn University)*; Jung Min Kim (University of Virginia); Bryan S Beckingham (Auburn University)

Poster 33 - Synthesis of Zinc Chloride Molten Salt Membranes for Potential Ammonia Separation
Lauren M Oleksy (Clarkson University)*; Moses Adejumo (Clarkson University); Simona Liguori (Clarkson University)

Poster 34 - Tunable nanofiltration and functionalized membranes for the separation of environmentally relevant ions
Francisco Leniz (University of Kentucky)*; Dibakar Bhattacharyya (U. Kentucky)

Poster 35 - Synthesis and characterization of ionenes. Preparation of membranes with possible applications in ionic transport.
Deysma Coll (Universidad Mayor)*; Alain Tundidor (Pontificia Universidad Catolica de Chile); Pablo Ortiz (Universidad Mayor)

Poster 36 - Transport Property Modulation via Non-Aqueous Solvent Specific Behavior in Cross Linked Membranes
Charles Leroux (University of Virginia)*; Patrick McCormack (University of Virginia); Gary Koenig (University of Virginia); Geoffrey M Geise (University of Virginia)

Poster 37 - Polyamide Reverse Osmosis Membrane Modification with Diamines to Improve Small Neutral Molecule Rejection
Shahriar Habib (The University of Alabama)*; Madison A Wilkins (The University of Alabama); Steven T Weinman (The University of Alabama)

Poster 38 - The Impact of dynamical correlations on salt transport in solvated ion exchange membranes
Nico Marioni (The University of Texas at Austin)*; Akhila Rajesh (The University of Texas at Austin); Zidan Zhang (The University of Texas at Austin); Harnoor Sachar (The University of Texas at Austin); Benny Freeman (The University of Texas at Austin); Venkat Ganesan (The University of Texas at Austin)

Poster 39 - The fundamental understanding of the effect of the H-bonds on CO_2/gas separation properties
Narjes Esmaeili (The State University of New York at Buffalo)*; Haiqing Lin (The State University of New York at Buffalo)

Poster 40 - Investigation of the impact of sidechain comonomer in the ion exchange membrane (IEM) on the electrochemical cell device performance.
Antara Mazumder (Auburn University)*; Bryan S Beckingham (Auburn University)

Poster 41 - Surface nanostructure control by the templateless electropolymerization
Imen Bousrih (University of Arkansas)*; Jamie Hestekin (University of Arkansas)

Poster 42 - 2D materials-based nanocomposite membranes for the removal and destruction of per- and polyfluorinated substances (PFAS)
Muhammad Usman Yousaf (University of Kentucky)*; Isabel Escobar (University of Kentucky)

Poster 43 - Specific Energy Analysis of Using Fertilizer-Based Liquid Desiccants for Indoor Plant Environments Dehumidification
Sarah Moussaddy (Oakland University)*; Jonathan Maisonneuve (Oakland University); Sandeep Aryal (Oakland University)

Poster 44 - Utilizing Greener Solvents for Membrane Fabrication
Joseph T Danner (The University of Alabama)*; Liam G Papier (The University of Alabama); Maggie L Callahan (The University of Alabama); Steven T Weinman (The University of Alabama)

Poster 45 - Ammonia Separation via Innovative Immobilized Molten Salt Membrane
Moses T Adejumo (Clarkson University); Lauren M Oleksy (Clarkson University); Simona Liguori (Clarkson University)*

Poster 46 - Multi-pass Nanofiltration for Lithium Separation with High Selectivity and Recovery
Ruoyu Wang (Vanderbilt university); Rayan Alghanayem (Vanderbilt University)*; Shihong Lin (Vanderbilt University)

Poster 47 - Development of Surface Patterned Super-oleophobic Poly (vinylidene fluoride) Membranes via Confined Thermally Induced Phase Separation for Enhanced Oil-Water Separation
Shouhong Fan (University of Colorado at Boulder)*; Duong T. Nguyen (University of Colorado Boulder); Kieran R Fung (Membrane Science, Engineering, and Technology (MAST) Center, Department of Mechanical Engineering, University of Colorado Boulder); Anthony Straub (University of Colorado Boulder); Yifu Ding (Membrane Science, Engineering, and Technology (MAST) Center, Department of Mechanical Engineering, University of Colorado Boulder)

Poster 48 - Experimental and Computational Evaluation of the Interaction Between Per- and Polyfluoroalkyl Substances with Microplastics in Aqueous Systems
Gabriel G Duarte Barbosa (The University of Alabama)*; Jasneet K Pala (The University of Alabama); Medha Kasula (The University of Alabama); Milad Esfahani (The University of Alabama); C Heath Turner (The University of Alabama)

Poster 49 - Effectively Using Heat to Thermally Enhance Reverse Osmosis and Reduce Specific Energy Consumption of Desalination
Sanjana Chintalacheruvu (Oakland University)*; Saber Khanmohammadi (Oakland University); Jonathan Maisonneuve (Oakland University)

Poster 50 - Improving the stability of supported ionic liquid membranes (SILMs) for the fractionation of light paraffins from natural gas streams
Justin J Rosenthal (The University of Texas at Austin)*

Poster 51 - Engineering a biocatalytic membrane for caffeine upcycling
Hadi Rouhi (University of Alabama)*; Meredith Mock (The University of Alabama); Ryan Summers (The University of Alabama); Steven T Weinman (The University of Alabama)

Poster 52 - Preventing mineral scale via alternating electric field for enhanced membrane-based and heat-exchanger-based processes
Yiming Liu (UCLA)*; Eric Hoek (UCLA); David Jassby (UCLA)

Poster 53 - Techno-Economic Analysis of Membrane-adsorption Hybrid Process for Direct Air Capture
Yi-Chen Huang (The Ohio State University)*; Yang Han (The Ohio State University); Winston Ho (The Ohio State University)

Poster 54 - Spacers for Mass Transfer Enhancement in Flat Sheet Membrane Modules – An Upper Bound?
Arsalan Sepehri (University of Toledo); Glenn Lipscomb (University of Toledo, Toledo)*

Poster 55 - Carbon capture demonstrated at ~1 tonne CO_2/day scale by an integrated membrane skid
Yang Han (The Ohio State University)*; Yutong Yang (The Ohio State University); Babul Prasad (The Ohio State University); Ruizhi Pang (The Ohio State University); Winston Ho (OSU)

Poster 56 - Creation of Polymer Datasets with Targeted Backbones for Screening of Gas Permeability and Selectivity
Surya P Tiwari (NETL, Battelle)*

Poster 57 - Development of a carbon capture membrane from an existing product and it's laboratory performance under highly realistic conditions
Max J Roberts (Compact Membrane Systems); Christine Parrish (Compact Membrane Systems)*; Ken Loprete (Compact Membrane Systems)

Poster 58 - x-ZAC Membranes for Ion Separations
Ashleigh B Herrera (Tufts University)*

Poster 59 - A new generation of expanded PTFE membranes for harsh chemical filtration
Zhefei Li (Pall corporation); Catherine Agcaoili (Pall corporation); Kyla White (Pall corporation); Timothy McKeon (Pall corporation); SIAVASH DARVISHMANESH (Pall corporation)*

Poster 60 - Biofouling Performance of Aromatic Polyamide Desalination Membranes Containing Functionalized Cellulose Nanocrystals (CNCs)
Connor L Farrell (Virginia Tech)*; E. Johan Foster (University of British Columbia); Stephen M Martin (Virginia Tech)

Poster 61 - Performance improvement of polyvinylidene fluoride (PVDF) nanocomposite membranes for highly saline water desalination using air gap membrane distillation (AGMD) process.
Toheeb O Obidara (King Fahd University of Petroleum & Minerals (KFUPM))*; Turki Nabieh Baroud (KFUPM); Abdul Azeem Mohammed (KFUPM); Dahiru Lawal (KFUPM)

Poster 62 - Electrospray 3D Printing to Unlock Customized Membranes
Mert Can Hacifazlioglu (University of Connecticut); Jeffrey R McCutcheon (University of Connecticut)*

Poster 63 - Sterically Hindered Amine Membranes for Selective H_2S/CO_2 Separation
Shraavya Rao (Ohio State University)*; Xuepeng Deng (The Ohio State University); Yang Han (The Ohio State University); Li-Chiang Lin (The Ohio State University); Winston Ho (The Ohio State University)

Poster 64 - In-Situ Dynamic 3D Characterization of HPRO Membrane Platform Under X-ray Tomography
Mi Zhang (University of Connecticut)*; Yara Suleiman (University of Connecticut); Jeffrey R McCutcheon (University of Connecticut); Sina Shahbazmohamadi (University of Connecticut); Mayur Ostwal (University of Connecticut)

Poster 65 - Decoding polymer behavior: The power of molecular modeling
Feranmi Olowookere (The University of Alabama)*; Heath Turner (The University of Alabama); Ali Al Alshaikh (The University of Alabama); Jason Bara (The University of Alabama)

Poster 66 - Mixed salt partitioning in model poly(ethylene oxide)-based membranes materials
Everett S Zofchak (The University of Texas at Austin)*; Aubrey Quigley (The University of Texas at Austin); Jordyn Yoh (The University of Texas at Austin); Venkat Ganesan (The University of Texas at Austin); Benny Freeman (The University of Texas)

Poster 67 - Hydrogen sulfide and acetylene stability and the effect of AgTf2N concentration on fractional free volume of hydrogen-stable Ag+-based facilitated transport membranes
Matthew N Davenport (The University of Texas at Austin)*; Tangqiumei Song (The University of Texas at Austin); Caitlin Bentley (The University of Texas at Austin); Taylor McClung (The University of Texas at Austin); Cara Doherty (CSIRO); Joan Brennecke (The University of Texas at Austin); Benny Freeman (The University of Texas at Austin)

Poster 68 - A Novel Method to Quantify Compaction of Desalination and Watertreatment Membranes
Weijian Ding (University of Southern California)*; Kexin Ma (University of Southern California); Shounak G Joshi (University of Southern California); Amy Childress (University of Southern California)

Poster 69 - Biocompatible cellulose-based phase inversion membrane to treat renal complications
Partha Pratim Chowdhury (University of Arkansas)*; Jamie Hestekin (University of Arkansas)

Poster 70 - MIL-100 (Fe) Functionalized Thin-film Nanocomposite Membranes for Enhanced Removal of Uncharged Urea from Water
Tin Le (The University of Alabama)*; Milad Rabbani Esfahani (The University of Alabama)

Poster 71 - Polyamide-based ion exchange membrane synthesis
Deniz Ipekci (University of Connecticut)*; Jeffrey R McCutcheon (University of Connecticut); Geoffrey M Geise (University of Virginia); Anna Harris (University of Virginia); Inara Oliviera (University of Virginia); Yasmin Teixeira Bimbatti (UCONN)

Poster 72 - Membrane fouling during the reclamation of industrial process water in a hybrid forward osmosis-reverse osmosis system
Catherine Murphy (Tennessee Tech University)*; Laura H Arias Chavez (Tennessee Tech University)

Poster 73 - Evaluating the Performance of Biochar-based Membranes Synthesized from Various Green Solvent for Application in Water Treatment
Abelline K Fionah (University of Kentucky)*; Aviana Judd (University of Kentucky); Isabel Escobar (University of Kentucky)

Poster 74 - Just Click 'Em Together: Controlling Swelling in New Polytriazole PIMs by Cross-Linking with the CuAAC Reaction
Valentino B Perez (Georgia Insitute of Technology)*; M.G. Finn (Georgia Institute of Technology); Youngjoo Lee (Georgia Institute of Technology); Ryan Lively (Georgia Tech)

Poster 75 - Assessing Suitability of Nanofiltration for Pretreatment of Lithium-Enriched Brines
Juan Zhai (Texas Tech University)*; Yuexiao Shen (Texas Tech University); Mahdi Malmali (Texas Tech University); rajesh khare (Texas Tech University); Amrika Deonarine (Texas Tech University); Shubhra Bhattacharjee (Texas Tech University); Adegbola Balogun (Texas Tech University)

Poster 76 - Upcycled Polyvinyl Chloride (PVC) Electrospun Nanofibers from Waste PVC-based Materials for Water Treatment
Atta Ur Razzaq (The University of Alabama)*; Milad Rabbani Esfahani (The University of Alabama)

Poster 77 - Electrospun Slow-Release Anti-Quorum Sensing Fibers Mitigate Biofouling in MCE Membranes
Amos Taiswa (Montana Tech)*; Jessica Andriolo (Montana Tech); Jack Skinner (Montana Tech)

Poster 78 - Defining a physically-relevant pore size for polyamide membranes
Rui Chen (Vanderbilt University)*

Poster 79 - Employing gutter layers on printed gas separation membranes to improve membrane permeance
Arjun DeJesus (University of Connecticut)*; Noah Ferguson (University of Connecticut); Srivatsa Bettahalli (University of Connecticut); Jeffrey R McCutcheon (University of Connecticut); Edward Wazer (University of Connecticut)

Poster 80 - Aggregation behavior and particle size populations: effect of ionic strength and cation choice
Kathlyn N Mealio (Tennessee Tech University)*; Martha Wells (EnviroChem Services); Holly Stretz (Tennessee Tech University)

Poster 81 - Characterizing liquid intrusion during membrane distillation: impacts of membrane properties, salinity, pressure, and fouling
Shounak G Joshi (University of Southern California)*; Allyson L McGaughey (Princeton University); Amy Childress (University of Southern California)

Poster 82 - Interfacial poly(epoxyether) membranes, a versatile thin-film synthesis platform for carbon capture membranes
Daan Van Havere (KU Leuven)*; Rhea Verbeke (KU Leuven); Benny Freeman (The University of Texas at Austin); Ivo Vankelecom (KU Leuven)

Poster 83 - Computational modeling of patterned membranes and spacers for improved hydrodynamics and fouling reduction in reverse osmosis water treatment processes
Zuo Zhou (Clemson University)*; David Ladner (Clemson)

Poster 84 - Carbon Molecular Sieve-derived POC/Mixed-Matrix Membranes for Gas Separation
Isha Kowey (Cleveland State University); Ali A Rownaghi (U.S. Department of Energy/Cleveland State University)*

Poster 85 - Nanoporous Atomically Thin Graphene Filters for Nanoscale Aerosols
Piran Kidambi (Vanderbilt University)*

Poster 86 - Acoustically actuated 3D printed bubble-trapping structures for in-situ mitigation of membrane fouling

Kieran R Fung (Membrane Science, Engineering, and Technology (MAST) Center, Department of Mechanical Engineering, University of Colorado Boulder)*; Shouhong Fan (University of Colorado at Boulder); Jason P. Killgore (Applied Chemicals and Materials Division, National Institute of Standards and Technology, Boulder, CO); Yifu Ding (Membrane Science, Engineering, and Technology (MAST) Center, Department of Mechanical Engineering, University of Colorado Boulder); Xiaoyun Ding (Department of Mechanical Engineering, University of Colorado Boulder)

Poster 87 - The Viability of Batch Reverse Osmosis with Pressure Retracted Osmosis

Sultan Alnajdi (Purdue University)*; Ali Naderi Beni (Purdue University); David M Warsinger (Purdue)

Poster 88 - Electrospray Synthesis of Highly Permselective Contorted Polyamide Membranes with Increased Fractional Free Volume

Parisa Taheri (University of Houston)*; Devin Shaffer (University of Houston)

Poster 89 - Enhancing Permselectivity of Polyamide Desalination Membranes by Incorporating Shape-Persistent Contorted Monomers

Siddhartha Paul (University of Houston)*; Sayali Shaligram (University of Houston); Rahul Shevate (University of Houston); Devin Shaffer (University of Houston)

Poster 90 - Pressure induced diffusion of solvents in dense homogeneous membranes

Kevin Reimund (University of Texas at Austin)*; Rahul Sujanani (University of California, Santa Barbara); Jorge Hernandez (University of Texas at Austin); Kristofer Gleason (University of Texas at Austin); Manish Kumar (University of Texas at Austin); Benny Freeman (The University of Texas at Austin)

Poster 91 - Bipolar membrane electrodialysis for direct LiOH production from concentrated brine

Aubrey Quigley (The University of Texas at Austin)*; Kevin Reimund (University of Texas at Austin); Nathaniel Lynd (The University of Texas at Austin); Benny Freeman (The University of Texas at Austin)

Poster 92 - Covalent Organic Framework (COF) Membranes Engineered for Selective Lithium-Ion Transport

Erin S Picton (University of Houston)*; Parisa Taheri (University of Houston); Devin Shaffer (University of Houston)

Poster 93 - Key Parameter Selection of Modified Phase Inversion GO-PSF Membranes for Membrane Distillation desalination

Lucy M Camacho (Texas A&M University-Kingsville, Department of Environmental Engineering)*; Olubayo O. Oginni (Texas A&M University-Kingsville, Department of Environmental Engineering); Kristin L. Trevino (Texas A&M University-Kingsville, Department of Environmental Engineering)

Poster 94 - Adsorptive Thin-film Nanocomposite Membranes Functionalized with Superfine Activated Carbon for PFAS Removal from Water

Medha Kasula (The University of Alabama)*; Milad Rabbani Esfahani (The University of Alabama); Jasneet K Pala (The University of Alabama)

Poster 95 - Removal of Micro- and Nano-plastics from Water using Liquid-Liquid Extraction

Ashish Srivastava (The University of Alabama)*; Milad Rabbani Esfahani (The University of Alabama); Jason Bara (University of Alabama).

Poster 96 - Ultrasonication Assisted Membrane Surface Functionalization by Silver (Ag)- and Zinc (ZIF-8) Metal-Organic Frameworks for Water Treatment: Synthesis, Characterization, and Performance

Himangshu Mondal (The University of Alabama)*; Medha Kasula (The University of Alabama); Milad Rabbani Esfahani (The University of Alabama)

Poster 97 - C18 functionalized nanodiamonds and MEEP-based membranes for efficient CO_2/N_2 separation

Christopher Orme (Idaho National Laboratory); Birendra Adhikari (Idaho National Laboratory)*; John Klaehn (Idaho National Laboratory); Joshua McNally (Idaho National Laboratory); Frederick Stewart (Idaho National Laboratory)

Poster 98 - Characterization and Associated Pressure-Dependent Behavior of Deposited mRNA-Lipid Nanoparticles Formed During Sterile Filtration

Kevork Oliver Messerian (The Pennsylvania State University)*

Poster 99 - Real-Time Induced Magnetic Vibrational Based Antifouling Mechanism for Ultrafiltration (UF) membrane

Jasneet K Pala (The University of Alabama)*; Ryan Tracy (The University of Alabama); Milad Rabbani Esfahani (The University of Alabama); Nima Mahmoodi (The University of Alabama)

Poster 100 - Tunable Poly(Ionic Liquid) Membranes for Gas Separations
Sudhir Ravula (University of Alabama); Kevin W Wise (University of Alabama)*; Jason Bara (University of Alabama)

Poster 101 - Correlation between volatile fatty acid (VFA) transport properties in anion exchange membranes and VFA transport during electrodialysis
Holly M Haflich (University of North Carolina- Chapel Hill)*; Joshua Singleton (University of North Carolina at Chapel Hill); Orlando Coronell (University of North Carolina at Chapel Hill)

Poster 102 - Synthesis of Peptoid Prepolymers for the Manufacture of Nanostructured Polyamide Membranes
Vladimir Niebla Morejón (Pontificia Universidad Catolica de Chile)*; Alain Tundidor (Pontificia Universidad Catolica de Chile); Claudio Terraza (Pontificia Universidad Catolica de Chile)

Poster 103 - Selective removal of selenium from saline water using different MOFs and MOF-incorporated NF membrane
Sweta A Modak (The University of Alabama)*; Milad Rabbani Esfahani (The University of Alabama)

Poster 104 - The Role of Membranes in Upcycling and Depolymerization of Plastic Wastes
Jason Bara (University of Alabama)*; Ali Alshaikh (University of Alabama); Mousumi Rani Bepari (University of Alabama, Tuscaloosa)

Poster 105 - Effect of Natural Organic Matter on PFOS Removal Using Adsorptive Thin-film Nanocomposite Membranes
Lela G Connors (The University of Alabama); Isabella Le (The University of Alabama); Medha Kasula (The University of Alabama); Milad Rabbani Esfahani (The University of Alabama); Kaitlyn Crowe (The University of Alabama)*

Poster 106 - Novel Upcycled Waste PVC Pipe Support Layer for Fabrication of Thin Film Composite Membranes for Water Treatment
Brenner Kar (The University of Alabama)*; Elizabeth Collinson (The University of Alabama); Ryan Hess (The University of Alabama); ATTA UR RAZZAQ (The University of Alabama); Milad Rabbani Esfahani (The University of Alabama)

Poster 107 - Wave Energy Converter system powering a batch reverse osmosis desalination system
Sultan Alnajdi (Purdue University)*; David M Warsinger (Purdue); Ali Naderi Beni (Purdue University); Mateo Roldan Carvajal (Purdue University)

Poster 108 - Degradation and separation of perfluorooctanoic acid (PFOA) from drinking water sources
Maria P Rincon (West Virginia University)*; Paola Perez Vega (West Virginia University); Oishi Sanyal (West Virginia University); John Hu (West Virginia University)

Poster 109 - Long-term disinfectants: The effects of cellulosic carriers on disinfectant materials
Eric A Walker (University of Arkansas)*

Poster 110 - Effect of Dehydration on RO Desalination Performance in Fully Aromatic Polyamide Membranes Containing TEMPO-Oxidized Cellulose Nanocrystals
Connor L Farrell (Virginia Tech)*; E. Johan Foster (University of British Columbia); Stephen M Martin (Virginia Tech)

Poster 111 - Forward Osmosis Oral Rehydration System Abstract
Elizabeth B Wilson (University of Arizona)*

Poster 112 - Understanding the effect of polymer morphology on the selective transport of ions
Meng Shen (California State University, Fullerton)*

Poster 113 - Use of Oxidized Micron Cellulose and Ionic Liquid Solutions to Cast Ultra-Filtration Membrane by Phase Inversion
Juliana Marcal (University of Arkansas); Jamie Hestekin (University of Arkansas)*

Poster 114 - Characterization of Poly(Arylene Ether)s for Membrane-Based Gas Separation
Duha Syar (MIT)*

Poster 115 - Understanding Water Absorption, Percolation, and Ion Transport in an Anion Exchange Membrane
Zhongyang Wang (University of Chicago)*

Poster 116 - Printed ultrathin film Nafion™ composite membranes by electrospray
Yasmin Teixeira Bimbatti (UCONN)*; Jeffrey R

McCutcheon (University of Connecticut)

Poster 117 - Poultry Slaughterhouse Wastewater Treated with Ultrafiltration and Reverse Osmosis for Reuse

Sana Fatima (Center of Energy and Environmental Sustaibaility)*; Hongbo Du (Prairie View A&M University); Raghava Kommalapati (Prairie View A&M University)

Poster 118 - High-Throughput Experimentation for High-Pressure Reverse-Osmosis Membranes

Hongchen Wang (University of Toronto); Jay Werber (University of Toronto)*

Poster 119 - Scaling-Up and Modelling the Fabrication of a More Sustainable Polymeric Membrane for Water Treatment Using Slot Die Coating

David Lu (University of Kentucky)*; Kwangjun Jung (Georgia Institute of Technology); Tequila Harris (Georgia Institute of Technology); Isabel Escobar (University of Kentucky)

Poster 120 - Energy reduction in ceramic microfiltration using supersaturated carbon-dioxide-enhanced backwash

Weiming Qi (Clemson University)*; David Ladner (Clemson)

Poster 121 - Synthesis of Tröger's base linked dibenzo-crown ether polymers and their use as gas separation membranes

Creighton Baltier (The University of Alabama)*; Sourav Chatterjee (Murray State University); Pravin Shinde (The University of Alabama); Kevin W Wise (University of Alabama); Jason Bara (University of Alabama)

Poster 122 - Investigation of the use of PolarClean as a green solvent alternative for PVDF membrane fabrication

Rachel Kaplan (University of Kentucky)*; David Lu (University of Kentucky); Isabel Escobar (University of Kentucky)

Poster 123 - Effect of PPE decontamination processes on porous polysulfone (PSf) flat sheet membranes fabricated via nonsolvent-induced phase separation (NIPS): A comparison with traditional PPE

Ebuka Ogbuoji (University of Kentucky)*; Isabel Escobar (University of Kentucky)

Poster 124 - Electrospun Membranes for Bioreactor Harvesting

Nimisha Tandon (University of Arkansas)*; Ranil Wickramasinghe (University of Arkansas); Xianghong Qian (University of Arkansas)

Poster 125 - Kinetic Control of Angstrom-scale Porosity in Graphene for Direct Scalable Synthesis of Atomically Thin Proton Exchange Membranes

Nicole Moehring (Vanderbilt University)*

Poster 126 - Surrogate-based Optimization for the Techno-economic Feasibility Analysis of Membrane Capture Chromatography Platforms

Juan Romero (Clemson University)*; Scott Husson (Clemson University); Eleanor Jenkins (Clemson University)

Abstracts – Oral Presentations, Monday

Presenting authors are indicated by an asterisk*.

Oral 1 – Polymer Design for Membranes

9:30 am (Druid, Monday)
Oral 1 - Versatile synthetic platform using click chemistry to study water and ion transport in post-functionalized membranes

Joshua D Moon (University of Florida)*; Oscar Nordness (University of California, Santa Barbara); Peter Richardson (University of California, Santa Barbara); Rahul Sujanani (University of California, Santa Barbara); Raphaële Clément (University of California, Santa Barbara); Benny Freeman (The University of Texas at Austin); Rachel Segalman (University of California, Santa Barbara); Craig Hawker (University of California, Santa Barbara)

Identifying new functional membrane materials with solute-tailored selectivity could enable recovery of solutes from non-traditional source waters, such as lithium from wastewater or brines, which is not possible with conventional materials. Unfortunately, identifying functional groups that could enable such precision separations is hindered by synthetic difficulties of efficiently incorporating diverse chemical moieties into well-controlled polymer structures.

10:00 am (Druid, Monday)
Oral 2 - Deep Eutectic Solvent (DES) – Inspired Design of Membranes for CO_2/Gas Separation

Narjes Esmaeili (The State University of New York at Buffalo)*; Haiqing Lin (The State University of New York at Buffalo)

Deep Eutectic Solvents (DESs) comprised of a hydrogen donor and a hydrogen acceptor have been demonstrated with enormous CO2 solubility and CO2/CH4 solubility selectivity. For example, a DES containing polyethylene glycol [PEG-8] and tetrabutylammonium bromide (TBAB) at 1 bar and 25 ℃ shows CO2 solubility of 7.2 cm3(STP)atm-1cm-3 and CO2/CH4 solubility selectivity of 46, much higher than those reported for any polymers. Inspired by the remarkable CO2 sorption in DESs, we design DES-mimic polymers and determine their potential for membrane CO2/gas separation. Specifically, poly(ethylene glycol) methacrylate (PEGMA) containing OH groups (H donor) is polymerized with different amounts of TBAB (H acceptor) and thoroughly characterized for physical properties and CO2/gas separation properties. Crystalline TBAB can be dissolved in the polymer (pPEGMA), and adding 20 mass% TBAB decreases glass transition temperature (Tg) from − 53 to − 60 ℃. By contrast, when the TBAB is added to cross-linked PEO without OH groups, TBAB cannot be fully dissociated by the polymer, confirming the H-bonding between pPEGMA and TBAB. C2H6 was used as a marker for the study of CO2 sorption, as both have similar condensability. Surprisingly, adding 20 mass% TBAB decreases CO2/C2H6 solubility selectivity from 3.6 to 2.6 at 35 ℃ because the ions interact with ether oxygen groups, which diminishes the strong affinity of CO2 to ether oxygen. Consequently, introducing TBAB in pPEGMA has minimal influence on CO2 permeability and CO2/N2 selectivity because of the competing effect of increased diffusivity (caused by the lower Tg) and decreased CO2 solubility. We have also studied pPEGMA-based DESs with other H acceptors, such as methyltriphenylphosphonium bromide (MTPB) and choline chloride (ChCl). The fundamental understanding of the effect of the H-bonds on CO2/N2 and CO2/CH4 separation properties can be useful to guide the design of the polymers for carbon capture.

10:30 am (Druid, Monday)
Oral 3 - High performance gas separation membranes from CANAL ladder polymers

Ashley Robinson (Stanford University)*

Polymers of intrinsic microporosity (PIMs) have emerged as exciting membrane materials for gas separations, defining the new performance upper bounds for polymer membranes. However, PIMs are predominantly limited to two types of rigid and contorted ladder-type backbone structures with ladder aryl ether or Tröger's Base linkages. We recently developed an efficient annulation polymerization (termed CANAL) to synthesize microporous ladder polymers with rigid norbornyl benzocyclobutene backbones. Tuning these polymers' chain configurations allows for dramatic enhancement of their membrane gas separation properties. With optimized design, we have obtained mechanically robust membranes with performance beyond the most recent upper bounds for many important gas pairs, with H2/CH4 selectivity >100 and CO2/CH4 selectivity >40. We investigated the structural requirements that lead to this exceptional separation performance by exploring polymers with systematic variations in chemical structures. This research provides new insight to the design of membrane polymer structures for the desired combination of high selectivity and high permeability for gas separations.

11:00 am (Druid, Monday)

Oral 4 - Upcycled Polyvinyl Chloride (PVC) Electrospun Nanofibers from Waste PVC-based Materials for Water Treatment

Atta Ur Razzaq (The University of Alabama)*; Milad Rabbani Esfahani (The University of Alabama)

Polyvinyl chloride (PVC) has grown to be one of the most popular types of plastic during the past few decades. However, PVC contaminates the environment and people throughout every stage of its lifecycle, from production to use and disposal. We report upcycling, the conversion of waste material to a high-value-added product, of PVC-based products to electrospun fibers (membranes). Two common PVC items, waste PVC pipe and waste PVC pool floats, were chosen and upcycled into electrospun membranes for water treatment. The fabrication process and fiber characteristics, such as morphology, surface charge, and mechanical strength of upcycled fibers, were studied and compared with the same fibers fabricated using research-grade (RG) PVC (commercial PVC powder). In addition, the effect of additives such as calcium carbonate in PVC waste products on the physicochemical properties of upcycled fibers was evaluated. The results showed that upcycling of waste PVC to electrospun fibers is feasible since the upcycled fibers showed similar or superior properties compared to their equivalent fibers from RG-PVC. Finally, the performance of upcycled fibers on the removal of dyes from the water was evaluated. The upcycled fibers from waste PVC pipe and pool float outperformed the RG-PVC fibers in removing methylene blue (MB) from water by showing more than 97% removal efficiency and adsorption capacity of 195 mg/g compared to RG-PVC with 89% removal and 179 mg/g adsorption capacity. This study clearly shows the potential of upcycling PVC waste materials from landfill into membranes for enhanced water treatment as a promising solution to the current plastic and water pollution.

11:30 am (Druid, Monday)

Oral 5 - Highly Permselective Contorted Polyamide Desalination Membranes

Devin Shaffer (University of Houston)*

The permeability-selectivity tradeoff observed for polymeric desalination membranes limits the performance of reverse osmosis desalination systems and increases their overall costs. Though membrane desalination, such as seawater reverse osmosis, has achieved high energy efficiency for separation, further improvements in membrane permeability and selectivity could intensify the desalination process, improving system energy efficiency, and reducing overall desalination costs. To intensify desalination system performance, the permeability-selectivity tradeoff of conventional polyamide desalination membranes, formed from the interfacial polymerization of m-phenylenediamine and trimesoyl chloride monomers, must be overcome. In this work, we demonstrate two new polyamide desalination membrane chemistries with water/NaCl permselectivities that exceed the current polyamide upper bound while maintaining NaCl rejection greater than 99%. These new polyamide membranes incorporate shape-persistent "contorted" diamine monomers into the polyamide network to increase network free volume. We fabricated contorted polyamide membranes using both a solution-based molecular layer-by-layer (mLbL) deposition technique and a more scalable electrospray synthesis technique. The chemistry, structure, and desalination performance of the contorted polyamide membranes will be presented and compared to that of conventional polyamide membranes. The results of diffusion-based transport modeling, that incorporates the measured and calculated fractional free volumes of the contorted polyamide materials, will be introduced to provide insight into the mechanisms of increased membrane permselectivity. The implications of this work for improving throughput, reducing costs, and expanding membrane desalination applications will be discussed.

12:00 pm (Druid, Monday)

Oral 6 - Understanding the Structure-Property Relationship of Poly(Ether Ether Ketone)-Ionene Composite Membranes for Gas Separations

Sudhir Ravula (University of Alabama)*; Kevin W Wise (University of Alabama); Jason Bara (University of Alabama)

Over the past decade, our group has been developing a library of ultrahigh-performance (UHP) and high-performance (HP) ionenes, a class of charged polymers where the ionic moieties reside within the polymeric backbone (as opposed to pendants). These UHP and HP-ionenes combine various functional groups associated with engineering polymers such as polyimides (e.g., Kapton) and poly(arylamides) (e.g., Kevlar, Nomex), which provide excellent mechanical properties for membrane formation, and the ability to hold additional "free" ionic liquids (ILs) within their structures. Currently, we are focused on synthesizing imidazolium ionenes containing aromatic ether-ether-ketone linkages inspired by poly(ether ether ketone) (PEEK), a well-known UHP engineering polymer, with a high melting temperature (335 oC) and very limited solubility in common organic solvents. In this approach, imidazolium-PEEK monomer was prepared via a simple and straightforward three-step synthetic route (nucleophilic aromatic substitution followed by Friedel-Crafts acylation and Ullman coupling). Finally, the EEK-containing bis(imidazole) was reacted with an alkylimidazolium/aryl dihalide compounds, providing PEEK-ionenes in good yields. This work further investigates the effects of spacing and sequencing of functional features and ionic groups incorporated along the main chain. The performance of these PEEK-ionene materials were enhanced when impregnated with "free" imidazolium-based IL, which serves as a non-covalent cross-link between the polymer chains and provide intramolecular interactions within the framework. The structure-property relationships of the newly developed PEEK-ionene composite membranes were extensively characterized by FT-IR, TGA, DSC, and XRD analysis. All the newly developed PEEK-composite composite membranes exhibited moderately high CO2 permeabilities with reasonable CO2/N2 and CO2/CH4 permselectivities.

Oral 2 – Industrial Membranes 1

9:30 am (Yellow Hammer, Monday)
Oral 7 - An Improved Electrodialysis Method for Amino Acid Production

Jian Xu (Bayer Crop Science)*

This study relates to an improved electrodialysis method for producing an amino acid from a salt of the amino acid utilizing a three-compartment bipolar membrane electrodialysis process. Bipolar membrane electrodialysis (BME) enables production of an organic acid from an organic salt by water splitting, which provides the protons for the acid formation. However, the three-compartment BME technology is mostly limited for strong acid production. Due to the nature of low ion conductivity and low solubility, weak acid such as amino acid production by this method is not favorable. In this study, an exogenous acid is added to the acid compartment to covert the amino acid to an acid salt which greatly enhance the ion conductivity and the overall current efficiency. In addition, the acid salt form can also keep the produced amino acid in solubilized form which is essential to achieve high product concentration for commercial process.

10:00 am (Yellow Hammer, Monday)
Oral 8 - Some Key Considerations for Developing New Industrial Membranes

John F Thompson (NL Chemical Technology)*; Norman Li (NL Chemical Technology)

During our many years of experience successfully developing and commercializing RO and NF membranes, we always follow several key considerations. Some of the following key factors will be discussed with actual case examples of important industrial membranes.

(1) Energy Considerations
- Does it save energy, or is it used primarily in a high-energy process?
- Example: Ultra-high pressure RO vs. Alternatives
- Example: Membranes for Minimal-Liquid or Zero-liquid Discharge Processes

(2) Environmental Impact
- Does it require environmentally hazardous or unsustainable ingredients?
- Example: Green solvents for membrane production

(3) Innovation
- Is it novel and innovative? Is the performance promising?
- Example: Bio-inspired or naturally derived products

(4) Failure-mode Analysis (Six Sigma)
- Is the membrane synthesis robust and practical?
- Is the product robust enough for field use?
- Example: Thin-film composite production methods

The future research directions of industrial membranes will also be discussed.

10:30 am (Yellow Hammer, Monday)
Oral 9 - Repairing the Membrane Innovation Pipeline: How to Reconnect Academics and Industry For more Impactful Membrane Research

Jeffrey R McCutcheon (University of Connecticut)*; Edward Wazer (University of Connecticut); Marianne Pemberton (University of Connecticut); Mayur Ostwal (University of Connecticut); Mi Zhang (University of Connecticut); Yara Suleiman@uconn.edu (University of Connecticut); Sina Shahbazmohamadi (University of Connecticut)

The membrane innovation pipeline serves to translate early stage research into commercially relevant technologies. In recent decades, this pipeline has fractured as a disconnect between academic research and industrial needs has grown. In this talk, we explore this disconnect, its root causes, how we can repair it, and the roles that startups play in bringing membrane technology to market. We discuss the membrane technology hype cycle, with a focus on desalination as an example, as another "valley of death" that membrane technologies must traverse to be successful. Consideration of historical membrane technology successes, tragic failures, and pivoting technology platforms will be included in the discussion.

11:00 am (Yellow Hammer, Monday)

Oral 10 - Designing a New Generation of Reverse Osmosis Membranes with Chemical Resistance

Sue J Mecham (NALA Membranes)*; Judy Riffle (NALA Membranes); Mikayla Armstrong (NALA Membranes); Andrew Tibbits (NALA Membranes)

Increasing water scarcity is driving efforts to apply reverse osmosis (RO) technology to treat more challenging water sources, and the notable absence of new membrane materials and products limits the impact of those efforts. Today's RO membranes are layered thin film composites (TFCs) comprised of an ~100-200-nm thick, non-porous, polyamide barrier layer atop a porous polymeric mechanical support with a textile backing. While these membranes demonstrate desirable performance with respect to flux and rejection, the inherently rough surface and poor chemical resistance results in fouling that drives energy and cleaning costs and reduces membrane lifetime. RO membranes based on materials that are resistant to oxidants, acids, and bases would offer opportunities to operate RO systems free from biofouling, enable separations of a wide range of contaminants, and be durable to aggressive cleaning solutions. Using patented precision sulfonated polysulfones instead of polyamides, NALA Membranes has developed new RO membranes to reduce the cost, complexity, and climate impact of RO operations. The chemical structure, hydrophilicity, and topology of the top barrier layer are critical. NALA's ionic membranes reject salt mixtures that are commonly found in both brackish water and seawater by both size exclusion and Donnan rejection and have the unique capability to reject monovalent ions even in the presence of multivalent salts, unlike previous membranes of this class. Their chemical durability, even against strong oxidants such as sodium hypochlorite and hypochlorous acid that are commonly used for disinfection and for cleaning, are outstanding. The new chemically resistant membranes can also be operated over a wide pH range providing additional resistance to cleaning chemicals and processes that are needed for applications such as wastewater reuse that contain a broad spectrum of contaminant types from differing source water.

11:30 am (Yellow Hammer, Monday)

Oral 11 - Scale-up and commercialization of a novel membrane chemistry

Christopher J Drover (ZwitterCo, Inc.)*

Zwitterionic copolymer membranes provide breakthrough capabilities for treating difficult industrial wastewaters. First introduced at the 2014 NAMS meeting by Dr. Ayse Asatekin, these revolutionary membranes have extraordinary fouling resistance and feature self-assembled pore structure in the ~1 nm range. In 2018, ZwitterCo, Inc. was formed to commercialize this technology platform. In this presentation, co-founder and CTO Christopher Drover will share technical and commercial learnings from the scale-up of this technology into the now commercially available Z-Clear Superfiltration product. Application data and case studies from several 'historically unfilterable' wastewater streams will be presented, including landfill leachate, meat & poultry wastewater, and other heavily contaminated industrial wastewater sources.

12:00 pm (Yellow Hammer, Monday)

Oral 12 - 3D-Printed Spacer Membrane Elements

CJ Kurth (Aqua Membranes)*

Since the development of spiral wound membrane elements in the 1960's, the construction of the feed channel has centered around an extruded mesh feed spacer sheet. This component both sets the channel height, creates flow instabilities to help increase mixing, and allows for the conveyance of retained materials out of the element. While the subject of a considerable amount of optimization, (thickness uniformity, strand angle optimization, softer materials, etc.) improvements to the element performance resulting from feed spacer developments have lagged that of other materials, most notably the membrane.

An alternate construction of the feed channel is to apply a pattern of features directly to the membrane surface. This concept was originally developed under funding from DARPA in the early 2000's. More recent advances in the field of additive manufacturing have allowed this technology to be scaled to commercially viable volumes and costs. By incorporating 3D-printed features on brackish water RO membranes, the conventional balance between feed spacer thickness and pressure drop has been significantly shifted providing for 8040 elements containing 25% more active area, exhibiting half the pressure-drop at the same flow rates, and more uniform flow distribution when compared with conventional elements.

This paper will discuss the construction and manufacturing process used to prepare such elements, testing to characterize performance and quantify concentration polarization as a function print pattern, and present field data from a full-scale installation running for 9 months.

Oral 3 – Membrane Synthesis and Casting 1

9:30 am (Black Warrior, Monday)

Oral 13 - Functional Hollow Fibers Made from Polymers of Intrinsic Microporosity (PIMs) for Membrane-Based Separations

Philippe Jean-Baptiste (Massachusetts Institute of Technology)*; Zachary Smith (MIT)

Polymer hollow fibers represent an important form factor for membranes for use in energy-efficient gas separations and organic solvent nanofiltration. These membranes are favored over traditional flat sheet membranes due to their increased surface-area-to-volume ratios on the scale of 103 m2m−3, as well as their greater mechanical stability. Despite their versatility, polymer membranes have limited industrial deployment because of their inherent permeability–selectivity trade-off defined by the Robeson upper bound. To overcome this challenge, polymers with inefficient packing structures and high fractional free volume known as polymers of intrinsic microporosity (PIMs) have been developed. Once formed into membranes, these polymers, specifically PIM-1 and CNSBF-TBTrip-I poly(arylene ether) (PAE) can be post-synthetically modified via solid-state functionalization by targeting their nitrile groups thereby increasing their gas separation performance. In this work, high molecular weight PIM-1 was synthesized at a multi-decagram scale for use in hollow fiber spinning. Once the ideal dope composition was determined through cloud point measurements, hollow fiber membranes were spun using a variety of conditions. To determine their chemical and structural stability, the spun fibers underwent solid-state functionalization to yield derivatives of PIM-1, including an amine-functional PIM (PIM-NH2), a tert-butoxycarbonyl-functional PIM (PIM-tBOC), and corresponding versions of PIM-tBOC that had the tBOC group removed through thermal or acid deprotection (PIM-deBOC). Of these fibers, the PIM-NH2, PIM-tBOC, and PIM-deBOC (thermal) samples were functionalized with no significant structural changes. PIM-deBOC (acid), however, exhibited a higher degree of pore collapse that may lead to a drop in permeability as a result. This work will enable PIM-1 and CNSBF-TBTrip-I PAE hollow fibers to be manufactured and functionalized while maintaining their performance when compared to films.

10:00 am (Black Warrior, Monday)

Oral 14 - Carbon Capture Membranes Based on Ultrathin Amorphous Polyethers Enabled by Nanoconfinement and Interfacial Engineering

Haiqing Lin (The State University of New York at Buffalo)*; Gengyi Zhang (The State University of New York at Buffalo); Vinh Bui (University at Buffalo)

Ultrathin-film composite (uTFC) membranes are a leading technology for post-combustion carbon capture, and one key challenge is to fabricate a defect-free selective layer (~100 nm) with superior CO2/N2 separation performance. Herein, we developed high-performance membranes based on semi-crystalline blends of amorphous poly(ethylene oxide) (aPEO) and 18-crown-6 (C6) using two nanoengineering strategies. First, as the aPEO/C6 is fabricated into thin films in membranes, the crystallinity decreases with decreasing thickness and completely disappears at 500 nm or below because of the nanoconfinement. Second, polydimethylsiloxane is chosen as the gutter layer between the porous support and selective layer, and its surface is modified with bio-adhesive polydopamine (PDA, <10 nm) with an affinity towards aPEO, enabling the formation of the thin, defect-free, amorphous aPEO/C6 layer. For example, the membrane with a 110-nm layer containing 40 mass% C6 in aPEO exhibits CO2 permeance of 2200 GPU (1 GPU = 10-6 cm3(STP) cm-2 s-1 cmHg-1) and CO2/N2 selectivity of 27 at 35 °C, surpassing Robeson's upper bound. This work shows that engineering at the nanoscale plays an important role in fabricating high-performance uTFC membranes for practical separations.

10:30 am (Black Warrior, Monday)

Oral 15 - Utilizing Greener Solvents for Membrane Fabrication

Joseph T Danner (The University of Alabama)*; Steven T Weinman (The University of Alabama)

Commercial microfiltration (MF) and ultrafiltration (UF) membrane fabrication methods utilize toxic, petroleum-based solvents such as N-methyl-2-pyrrolidone (NMP) and N,N-dimethylformamide (DMF). Greener solvents are inherently less toxic, more environmentally friendly, and/or derived from renewable materials. By developing strategies to incorporate greener solvents into membrane fabrication, further societal and environmental impacts can be mitigated while maintaining the separation performance of current commercial membranes. This study investigates the structure-property-performance relationship of solvents and how they impact fundamental properties of polyethersulfone (PES) membranes prepared using nonsolvent-induced phase separation (NIPS). Solvents were selected based on the evaluation of their Hansen solubility parameters, toxicity, and partition coefficient (LogP). The membranes were prepared by the NIPS method using PES in two categories of solvents: conventional and greener. N,N-dimethylacetamide (DMAc) was used as the conventional solvent while dimethyl sulfoxide (DMSO) and 2-methylpyrazine (2MP) were used as the greener solvents. Cloud point titration was used to determine the thermodynamic stability of polymer solutions. Dead-end filtration was used to evaluate the pure water permeance and solute rejection of the membranes. Physical characterization of each membrane's cross-section and surface morphology were accomplished via scanning electron microscopy. Analysis of SEM images found that 2MP lead to an increased top surface pore size and a denser cross-sectional morphology. This denser cross-sectional morphology resulted in less compaction and lower permeances when operating the membranes at 3 – 4 bar.

11:00 am (Black Warrior, Monday)

Oral 16 - Combining Macrostructure, Microstructure and Molecular Scale Morphology to Create New High-Performance Composite Carbon Molecular Sieve Membranes

Yuhe Cao (Georgia Institute of Technology)*; Zhongyun Liu (Georgia Institute of Technology); Wulin Qiu (Georgia Institute of technology); William Koros (Georgia Institute of Technology)

Carbon molecular sieve (CMS) membranes are known for extraordinary intrinsic permeability and selectivity that greatly exceed the polymer upper bound trade-off curves for many gas pairs. Such intrinsic properties reflect structures present at the molecular scale of CMS materials. The added ability to maintain such intrinsic properties in an asymmetric morphology makes CMS materials unusual in terms of scalability to advanced hollow fibers and modules. Here we extend such properties by introducing asymmetric hollow fibers with a structure that we term a "Pseudo Wheel+ Hub & Spoke" asymmetric form. The novel structure provides needed selective layer support while minimizing porous flow resistance in the support for applications with potentially very high feed pressures and fluxes. Combining precursor asymmetric hollow fiber formation with optimized pyrolysis to create a defect-free CMS will be described for the important natural gas separation application. A proof-of-concept membrane useful for this application will be discussed and analyzed, stressing the much higher modulus of carbon vs. even robust glassy polymers. Specifically, we report a facile scalable method to fabricate composite CMS membranes from a sheath-core spun precursor with 6FDA-DAM sheath and Matrimid® core. The composite CMS membrane fabrication process avoids posttreatment to seal defects before or after the carbonization of the precursors. The composite CMS membrane shows CO2/CH4 (50:50) mixed gas feed with an attractive CO2/CH4 selectivity of 64.3 and CO2 permeance of 232 GPU at 35 °C. An additional benefit of the approach is a reduction in the use of the more costly high-performance 6FDA-DAM in a composite sheath-core CMS membrane with the "Pseudo Wheel+ Hub & Spoke" structure.

11:30 am (Black Warrior, Monday)

Oral 17 - Highly selective interfacial polymerized polyamide TFC membranes by tunning synthesis parameters

Chenhao Yao (University of Texas at Austin)*; Manish Kumar (University of Texas at Austin)

Polyamide (PA) thin film composite (TFC) membranes made from interfacial polymerization (IP), featuring a highly crosslinked polymer network, is widely applied in water purification processes. The rapid fabrication procedure and simple chemistry has made PA TFC membranes successful in commercial nanofiltration and reverse osmosis applications. Despite the outstanding desalination performance, PA TFC membranes do no show strong size-sieving ability at angstrom-scale and are not commercialized in gas separation applications. This is presumably due to the lack of control over the microstructure and crosslinking during the rapid IP procedure. In this work, the synthesis parameters of the TFC membranes were tuned to control the free volume formation and polymer crosslinking, and membranes with great size-sieving ability and gas separation performance were produced. TFC membranes were synthesized with a series of diamine monomers with distinct structural and functional characteristics, and the IP reaction was carried out with different monomer concentrations. The accessible free volume and surface area of the polymer matrix was simulated using MD simulation. In addition to the conventional m-phenylene diamine (MPD), four aliphatic diamines (Piperazine (PIPz), 4-amino Piperidine (PIPd), trans-diamocyclohexane (DAC) and bis-aminomethylcyclohexane (BAMC)) were used and the resultant membranes showed different free volume distribution according to the physiochemical properties of the monomers. By tuning the starting monomer concentration, the crosslinking of polymer was optimized, and the membranes showed H2/N2 selectivity up 537 and H2/CO2 selectivity up to 53.5 which exceeds the 2008 Robeson upper bound. This work proves that the microstructure and crosslinking of interfacial polymerized PA TFC membranes can be controlled by tuning the synthesis parameters, and the size-sieving ability of the membranes can be optimized towards commercial gas separation applications.

12:00 pm (Black Warrior, Monday)
Oral 18 - Electrospray Printed Membranes for Gas Separations

Noah Ferguson (University of Connecticut)*; Srivatsa Bettahalli (University of Connecticut); Mayur Ostwal (University of Connecticut); Jeffrey R McCutcheon (University of Connecticut)

Anaerobic digestion of organic wastes can provide a wealth of high value products including fertilizer, volatile fatty acids, and biogas. Upgraded to biomethane through the removal of CO2 and other impurities, biogas presents a renewable alternative to natural gas which could assist humanity's shift away from fossil energy. However, current biogas upgrading technologies are complex, expensive, and energy intensive. Here we present gas separation membranes made with electrospray deposition. Electrospray technology employs an electric field between the spray tip and the deposition surface. Charged spray droplets experience Coulombic fission during flight, generating a fine spray mist that can create membranes only tens of nanometers thick. Such membranes can be made from many solution-processable polymers, scalably and repeatably. Our present research employs Pebax-1657, a poly(ether block amide) copolymer, based on its demonstrated efficacy in CO2 gas separations. Pebax exploits its polyamide (PA) units to provide mechanical strength and prevent polyethylene oxide (PEO) crystallinity. PEO's dipole-quadrupole interactions increase CO2 membrane solubility, while its chain flexibility improves permeability. Because Pebax-1657 is a thermoplastic, membrane preparation is facile and avoids the need for crosslinking. Electrosprayed Pebax-1657 membranes demonstrate similar CO2/N2 selectivities and superior permeances compared to traditionally produced Pebax-1657 membranes in the literature. Thin, high permeance electrosprayed membranes may make biogas upgrading cheaper and simpler, and help establish membranes as the technology of choice for biogas treatment.

Oral 4 – Module Modeling and Design

9:30 am (Private Dining, Monday)
Oral 19 - Modeling gas separation in flat sheet membrane modules: Impact of flow channel size variation

Joanna Rivero (University of Pittsburgh)*; Glenn Lipscomb (University of Toledo, Toledo); Katherine Hornbostel (University of Pittsburgh); Leo Nemetz (University of Toledo); Marcos Da Conceicao Acosta (University of Toledo)

Flat sheet membranes offer many advantages over other membrane configurations, (e.g. ease of maintenance and low pressure drops) that make them a strong candidate for post-combustion carbon capture. A performance model for a stacked flat sheet membrane module is reported in this work. The model is based on the reported specifications for the Gen 2 Polaris™ module developed by Membrane Technology & Research (MTR) and predicts performance based on solution of the governing momentum and mass balance equations. The model accounts for variability in flow channel heights within the module that can arise during module manufacturing. The model is first verified against similar membrane models. Simulations are then performed over a wide range of conditions to demonstrate how much performance declines as channel height variability increases. As per the performance metrics, the dimensionless feed flow rate processed per unit membrane area (f-curve) shows the greatest decline. The changes in performance are comparable to those that occur with hollow fiber membrane modules that possess similar fiber size variations. Together, the results of this study indicate that flow channel height variation in flat sheet membrane modules can hurt CO2 separation performance, but the impact is minor except at low CO2 retentate compositions with large channel height variation (e.g. a 10% decline in stage cut performance for a mixture with 94% CO2 in the permeate stream and 30% channel height variation). However, high variation has a significant impact on overall membrane area with a 30% increase at 30% variation.

10:00 am (Private Dining, Monday)

Oral 20 - Hollow Fiber Membrane Modules for Flue Gas Separation: Experimental Verification of CFD Models using 3D Printing

Victor A Kusuma (NETL)*; Lili Sun (University of Toledo); Comfort Oluleke (University of Toledo); Nathan Diemler (NETL); Lingxiang Zhu (National Energy Technology Laboratory); Grigorios Panagakos (Carnegie Mellon University); Glenn Lipscomb (University of Toledo, Toledo); David Hopkinson (NETL)

Most of the recent research on gas separation membranes for CO2 removal from flue gas, an application with large societal importance, was aimed at making more permeable and/or more selective membranes to improve separation efficiency. To achieve true carbon capture cost reduction, however, it is equally important to evaluate the current module designs and optimize them for flue gas treatment. The low driving force due to low CO2 partial pressure in flue gases can be exacerbated by poor module design that leads to inefficient flow distribution and stagnant zones.
In this study, 3D printing using stereolithography was utilized to produce laboratory-scale hollow fiber membrane modules in parallel with a full module geometry computational fluid dynamics (CFD) simulation on the same model using COMSOL Multiphysics. This allows for a model performance prediction which can be corroborated using actual mixed gas permeation experiments. The 3D-printed modules can also be designed using flow configurations that cannot be achieved using compression fittings. Hollow fiber composite membranes comprised of a microporous poly(vinylidene fluoride) (PVDF) support, a polydimethylsiloxane gutter layer, and a Pebax 2533 selective layer were sealed and aligned within the module casing in different geometrically-defined arrays using buttons. Mixed gas experiments were performed using CO2/N2 mixtures that represent typical carbon capture applications in a counter-current configuration using a permeate vacuum.
The experiments showed the performance, as measured by product purity and flow efficiency, was highly influenced by feed flowrate and changes to the module geometry. For instance, moving the feed and retentate ports closer leads to stagnant flow zones and reductions in performance metrics. CFD modeling captured these results well: the experimental and modeling results are in good agreement.

10:30 am (Private Dining, Monday)

Oral 21 - Harnessing gravity to mitigate polarization and increase permeate production in membrane distillation

Nils Tilton (Colorado School of Mines)*; Federico Municchi (Colorado School of Mines); Yiming Liu (UCLA); Tzahi Cath (Colorado School of Mines); Michael Heeley (Colorado School of Mines); Eric Hoek (UCLA); Craig S Turchi (NREL); David Jassby (UCLA)

Membrane Distillation (MD) has been studied extensively for applications to desalinating complex wastewaters and hypersaline brines. MD nevertheless faces two persistent technological challenges. The fist, called temperature polarization, is the cooling of the feed in a thermal boundary layer near the membrane. This reduces the local rate of permeate production. The second, called concentration polarization, is the accumulation of solutes near the membrane, where they can precipitate. Surprisingly, no prior work considers that temperature and concentration polarization also increase the local feed density near the membrane. We present computational fluid dynamics (CFD) simulations that show that with gravity properly oriented, this increase in feed density can trigger buoyancy-driven convection, in which plumes of cool solute-rich feed sink away from the membrane. This in turn brings warm low-concentration feed to the membrane, mitigating temperature and concentration polarization. This mixing dramatically increases vapor production. We explore the dependence of buoyancy-driven convection on the operating and feed conditions, and we outline how it can be sustained over long membrane surfaces using external heating.

11:00 am (Private Dining, Monday)

Oral 22 - Demonstration of Direct Ocean Carbon Capture Using Membrane Technologies

Joanna Rivero (University of Pittsburgh)*; Austin R Lieber (University of Pittsburgh); Chris Snodgrass (University of Pittsburgh); Zoe Neal (University of Pittsburgh); Donna Hildebrandt (University of Pittsburgh); William Gamble (University of Pittsburgh); Shanna Davidson (University of Pittsburgh); Huda Usman (University of Pittsburgh); Tagbo Niepa (University of Pittsburgh)

The focus of membrane carbon capture to date has been primarily on point source capture, such as power plants and industrial capture. However, membrane technology can also play a role in negative emissions technology, such as direct air capture and direct ocean capture. Direct ocean capture has a few potential advantages over direct air capture, such as avoiding land use and coupling with offshore wind and offshore storage. The use and feasibility of hollow fiber membrane contactors (HFMCs) and encapsulated solvents for direct ocean carbon capture are assessed here through a multifaceted approach: modeling and experimentation. Separate 1D models for each technology incorporate fluid dynamics and the chemical kinetics of both ocean water and the respective aqueous solvent to simulate CO2 flux. Lab scale experiments of this system then guide a model refinement and validation process. Additionally, a preliminary techno-economic assessment (TEA) is performed based on the computational and experimental results to estimate the carbon capture cost when the system is scaled to remove 0.98 Mtonnes CO2/year. Computational results suggest that higher seawater flow rates and temperatures relative to the solvent for the HFMC improve CO2 flux. As for the encapsulated solvent, optimal flux was observed when the capsule diameter is between 400-600 μm for a balance between pressure drop and breakthrough time and porosity is between 0.5-0.7. The TEA estimates that these membrane technologies may only be cost-competitive if seawater pH is decreased at the membrane interface, thereby increasing the local concentration of dissolved carbon dioxide. These findings indicate that local pH swing on the seawater side will be necessary to feasibly remove carbon dioxide from seawater using HFMCs.

11:30 am (Private Dining, Monday)

Oral 23 - Energy and exergy analysis of multi-stage vacuum membrane distillation integrated with mechanical vapor compression

Mahdi Malmali (Texas Tech University)*; Bosong Lin (Texas Tech University)

Membrane distillation (MD) is a promising candidate for desalinating hypersaline brine, but its poor energy efficiency has remained a major barrier for widespread application. One possible solution to resolve this issue is to retrieve and reuse the latent heat in the process. In this work, a multi-stage vacuum MD (MSVMD) was integrated with a mechanical vapor compressor (MVC) to enhance the latent heat recovery, and the energetic and exergetic performance of this integrated process was examined. A comprehensive energy and exergy analysis will be presented to compare MSVMD and MSVMD-MVC processes for desalination of hypersaline brine. This analysis was conducted by examining the effect of the compression ratio on the energetic and exergetic performance, and the findings are reported in terms of specific thermal energy consumption (STEC), specific electrical energy consumption (SEEC), and exergetic efficiency. The energy analysis shows that thermal energy consumption can be reduced as the compression ratio increases, due to the enhancement of latent heat recovery. It was found that MSVMD-MVC process can be operated in steady state condition, without the need for thermal heat input; with the feed temperature of 50 °C and MVC compression ratio of 2.14, STEC and SEEC were found to be 0 and 49 kWh/m3. Moreover, exergy analysis demonstrated the efficacy of the eNRTL model in exergy calculation. Exergy destruction can be greatly reduced with increasing the MVC's compression ratio to an optimum value. The MVC-enhanced process achieved a higher exergetic efficiency of 6.85%, compared to 2.42% in the MSVMD process. The result suggests that the application of MVC can intensify the energy efficiency and exergetic efficiency of the MSVMD system, although this process cannot outperform the current desalination technologies from the standard primary energy point of view.

12:00 pm (Private Dining, Monday)

Oral 24 - A new approach to simulating concentration polarization in reverse osmosis systems with feed spacers

Nils Tilton (Colorado School of Mines)*; Jacob Johnston (Colorado School of Mines); Sarah Dischinger (LBNL); Mostafa Nassr (UT Austin); Ji Yeon Li (LBNL); Sergi Molins (Lawrence Berkeley National Laboratory); Nicolas Spycher (LBNL); Benny Freeman (University of Texas Austin); William Stringfellow (LBNL); Daniel Miller (LBNL); Kristofer Gleason (University of Texas Austin)

Feed spacers in reverse osmosis systems generate complex fluid flows that limit computational fluid dynamics (CFD) simulations to small length and time scales. That limits our ability to simulate mineral scaling and other membrane fouling phenomena, which occur over longer length and time scales. Thus motivated, we develop a reduced model that replaces the CFD simulation of the velocity field with an analytical model that mimics spacers. This focuses the remaining numerical effort on simulating the advection-diffusion equation governing solute transport. We motivate and validate the model with CFD simulations and bench-scale experiments of spacer filaments in three different arrangements, including cases of unsteady vortex shedding. We show that the model produces a roughly 10,000-fold speedup compared to CFD, and accurately reproduces CFD predictions of not only the average and maximum concentrations, but also the local concentration distribution along the membrane. The model provides a simple testbed for exploratory studies of multispecies transport, precipitation, and membrane fouling phenomena for which simulating spacers is often prohibitive.

Oral 5 – Carbon Capture and Facilitated Transport

9:30 am (Capstone, Monday)
Oral 25 - Scale-up of CO2-selective nanoporous single-layer graphene membranes

Jian Hao (Ecole polytechnique fédérale de Lausanne)*; Kumar Varoon Agrawal (Ecole polytechnique fédérale de Lausanne)

Atom-thick gas-sieving nanoporous single-layer graphene (NSLG) membranes have shown great potential in post-combustion carbon capture. CO2 permeance of above 10000 GPU and CO2/N2 separation factor near 30 from NSLG membranes was recently achieved by our group1,2, and a competitive capture penalty (30 to 40 $/tonCO2) for the case of post-combustion capture was assessed by a technoeconomic analysis3. However, the current high-performance graphene membrane is still minuscule, and scalable fabrication routes are needed. The most predominant factor that restricts the upscaling of the graphene membrane is the crack formation during the transfer of graphene film from metal foil to a porous support. In this presentation, we will discuss several interventions that we carried out to realize large-area graphene membranes. We adopted the crystallization protocol of graphene to a low-cost Cu foil. We used acid pre-treatment on Cu foil to remove contamination particles that are detrimental to the graphene membrane. The acid treatment in the acidic solution was found to be effective even for low-cost Cu foils. High-quality 0.3 m x 0.1 m graphene coupons could be synthesized with a low density of intrinsic defects. The graphene film was mechanically reinforced with a nanoporous layer and was transferred on a low-cost stainless-steel mesh while avoiding cracks and tears. The membrane was assembled inside a customized module and could be sealed without any leaks. The graphene films without any intentional pore generation step showed a low gas permeance confirming a low intrinsic defect density from graphene and crack-free membrane preparation. Incorporating graphene with CO2-selective pores resulted in attractive CO2-sieving performance. Our work paves the way to up-scaling graphene membranes and achieving a competitive carbon capture cost.

10:00 am (Capstone, Monday)
Oral 26 - Extending the effects of crosslinks during free volume manipulation (FVM) of amine-functionalized PIM-1 membranes for gas separation

Taigyu Joo (Massachusetts Institute of Technology)*; Sungsool Wi (National High Magnetic Field Laboratory); Zachary Smith (MIT)

Functionalization of polymer membranes with polar moieties like amines is a rational way of increasing CO2-based sorption selectivity. However, this approach may also reduce free volume by introducing secondary interactions like hydrogen bonds, making it difficult to increase CO2 sorption without compromising diffusivity. Free volume manipulation (FVM) is a post-synthetic solid-state modification strategy developed to increase the free volume of polymer membranes containing polar functionalities, overcoming the concomitant loss of free volume induced by the polar groups. FVM uses labile and bulky functional groups, such as a tert-butoxycarbonyl group, to intentionally widen intersegmental distances. When the labile groups are deprotected from the polymer using simple heat treatments, free volume is increased. This approach was demonstrated to be highly effective for amine-functionalized PIM-1 (PIM-NH2), which resulted in a simultaneous boost in permeability and selectivity. The light urea crosslinks that form during FVM were found to play a significant role in preserving the increased free volume, narrowing the free volume element distribution, and providing high energetic barriers for activated perturbation of polymer segments for the diffusion of penetrant molecules. In this study, the effects of crosslinks during FVM of PIM-NH2 are exploited further by systematically introducing a small amount of oxygen during the FVM procedure, inducing thermally oxidative crosslinks. The gas transport properties significantly changed with varying oxygen concentration, yielding an almost 10-fold increase in CO2/CH4 selectivity while losing only about 30% of CO2 permeability. In addition, the increased number of crosslinks that covalently bond polymer segments together provided a high resistance to CO2 plasticization. This study demonstrates a simple yet effective strategy to tune the diffusion properties of amine-functionalized polymers and yield stable membrane performance.

10:30 am (Capstone, Monday)
Oral 27 - Graphene Oxide/Single Walled Carbon Nanotube (GO-SWCNT) Network Restricted Ionic Liquid as Stable and Effective Membrane for Highly Efficient Carbon Capture

Dinesh K Behera (University at Buffalo)*; Bratin Sengupta (University at Buffalo); Fan Wang (University at Buffalo); Shiguang Li (Gas Technology Institute); Miao Yu (University at Buffalo)

Carbon capture, utilization and storage is considered to be one of the key strategies on alleviating climate change and promoting net removal of carbon dioxide (CO2) from atmosphere. A significant fraction of CO2 released into the atmosphere is from flue gas emissions in power plants. Membrane processes may offer a promising solution for low cost, robust and energy efficient CO2 capture from these plants. Facilitated transport membranes, although showing strong capability for CO2 capture at low concentrations and under humid conditions, suffer from the loss of facilitators leading to decline in performance over long-term operation. Ionic liquids (ILs) are chemically, thermally stable and have low vapor pressure, which is ideal to prevent aforementioned losses. However, mechanical stability of supported IL membranes, especially under pressure drop conditions, is a concern. In this work, we report the use of an amino acid IL, 1-ethyl-3-methylimidazolium glycinate ([EMIM] [GLY]), as the facilitator for CO2 capture. In order to selectively load and confine [EMIM] [GLY] in a thin selective layer, we fabricated a Single Walled Carbon Nanotubes (SWCNT) intercalated Graphene Oxide (GO) layer. The distribution of [EMIM] [GLY] in the GO-SWCNT network is expected to be critical for the formation of a thin selective layer. To understand and optimize this, the ratio of GO to SWCNT was varied in an attempt to adjust the interaction between the network and [EMIM] [GLY]. The negatively charged GO acts as anchoring sites for [EMIM] [GLY] molecules (due to electrostatic interaction between anions on GO and cations of IL), while the interconnected GO/SWCNT network provides numerous nanochannels for low-resistance gas transport. The influence of IL concentration, temperature, pressure and water on performance of the [EMIM] [GLY] restricted in GO/SWCNT framework was evaluated. The optimized membrane showed a CO2 permeance of >1,500 GPU and CO2/N2 selectivity >300 for simulated flue gas.

11:00 am (Capstone, Monday)

Oral 28 - Effect of Amine Content on the Carbon Dioxide Separation Performance of Polyvinylamine Membranes for Flue Gas Carbon Capture

James S Baker (National Energy Technology Laboratory)*; Victor A Kusuma (NETL); Zi Tong (National Energy Technology Laboratory); David Hopkinson (NETL)

Polyvinylamine (PVAm) is a polymer that has commonly been studied for carbon dioxide (CO2) separation facilitated transport membranes due to its high concentration of amine sites, ease of synthesis, and excellent film forming properties. PVAm is usually a random copolymer prepared by partial hydrolysis of poly(N-vinylformamide) (PNVf). Since the vinylamine units act as CO2 carriers for facilitated transport while the N-vinylformamide units do not, the permeability and selectivity of PVAm polymers is influenced by the fraction of sites that are amines. However, the relationship between amine content in the polymer and CO2 permeability and selectivity is not straightforward.

In this study, PNVf was hydrolyzed to various amine contents and the resulting CO2 permeability and CO2/N2 selectivity of bulk PVAm films with humidified 14/86 CO2/N2 mixed gas was measured. Initially, increasing the amine fraction in the polymer improved CO2 permeability and selectivity until about 60 mol% amine fraction, after which selectivity plateaued and permeability declined with increasing amine fraction. X-ray diffraction and humid dynamic mechanical analysis experiments showed PVAm was an amorphous polymer across the range of compositions; therefore, the decline in permeability was not due to crystalline domains impeding gas permeance. Instead the decrease was due to polyelectrolyte complexation of ammonium cations and carbamate anions covalently bound to the polymer chains when PVAm reacts with CO2 in the presence of water. When PVAm solutions were stirred under CO2 atmosphere, polymers having >60% amine content formed precipitates that were due to formation of these polyelectrolyte complexes. Because these inter-chain interactions impede gas diffusion, polymer membranes having a moderate amine content can yield higher CO2 permeability than those with high amine contents.

11:30 am (Capstone, Monday)

Oral 29 - Carbon capture from flue gases using an integrated membrane skid at ~1 tonne CO2/day scale

Yang Han (The Ohio State University)*; Yutong Yang (The Ohio State University); Babul Prasad (The Ohio State University); Ruizhi Pang (The Ohio State University); Winston Ho (OSU)

Large-scale application of membrane in decarbonizing flue gas has been limited by the insufficient CO2/N2 selectivity of most polymeric materials and the unproven thin membrane fabrication at scale. Aiming to overcome these limitations, a novel facilitated transport membrane (FTM) containing polyguanidine as fixed-site carrier was synthesized in a thin-film composite configuration. The FTM was also scaled up to 20" in width by a roll-to-roll process. A prototype spiral-wound (SW) membrane module was fabricated with a commercial-size diameter of 8" and a membrane area of 35 m2. This SW module was tested with an actual coal flue gas at the Center for Applied Energy Research (CAER) at the University of Kentucky, which exhibited an average CO2 permeance of 4269 GPU with a CO2/N2 selectivity of 165 and remained stable for 100 h.

An integrated membrane skid was then constructed, where 2 SW membrane modules were arranged to form an enriching cascade. The 8"-diameter prototype SW module previously tested at CAER was installed as the primary CO2 capture stage, while a smaller SW module (5" diameter and 12 m2 membrane area) was used to further enrich the CO2 to >95% purity. By using a simulated coal flue gas (13% CO2), the skid was operated at steady-state for 800 h with 91.0% of the CO2 removed from the flue gas, delivering 1.33 tonne/day CO2 product at 95.5% dry purity.

The bench skid was then installed at the National Carbon Capture Center (NCCC), Wilsonville, AL and tested with an actual natural gas (NG) flue gas (8.6% CO2). 90–99% CO2 capture degrees were demonstrated during the parametric testing, and the skid was operated at steady state for 500 h cumulatively with 91.0% CO2 capture and >95% CO2 purity. A simulated natural gas combined cycle (NGCC) flue gas (4.3% CO2) slipstream was further provided by diluting the NG flue gas with air, and 90–99% CO2 capture degrees were also achieved with dry CO2 purities all above 95%.

12:00 pm (Capstone, Monday)

Oral 30 - Hydrogen sulfide and acetylene stability and the effect of AgTf2N concentration on fractional free volume of hydrogen-stable Ag+-based facilitated transport membranes

Matthew N Davenport (The University of Texas at Austin)*; Tangqiumei Song (The University of Texas at Austin); Caitlin Bentley (The University of Texas at Austin); Taylor McClung (The University of Texas at Austin); Cara Doherty (CSIRO); Joan Brennecke (The University of Texas at Austin); Benny Freeman (The University of Texas at Austin)

Previous studies on Ag+-containing facilitated transport membranes report high olefin-paraffin selectivity (e.g., ethylene over ethane) but poor stability of the Ag+ carrier, especially in the presence of reducing gases, such as H2, H2S, and C2H2 which are often present in industrial process streams. Solid polymer electrolytes consisting of crosslinked poly(ethylene glycol) diacrylate and up to 70 wt% silver bis(trifluoromethylsulfonyl)imide (AgTf2N) salt are synthesized both as dense films and thin film composite membranes featuring an Ag+-containing active layer less than 1 µm thick. Following over 10 weeks of pure H2 permeation at 4 bar and 35°C, the membrane retains over 90% of its initial pure gas ethylene-ethane selectivity, performance which surpasses the polymeric gas separation membrane upper bound. The membrane is similarly resistant to C2H2 exposure, however altered transport properties and formation of Ag2S was observed following H2S exposure. Furthermore, Positron Annihilation Lifetime Spectroscopy (PALS) was used to characterize the free volume of the facilitated transport membranes as a function of AgTf2N concentration, showing that the fractional free volume decreases non-linearly as a function of salt concentration. This result is discussed in the context of polymer chain dynamics (Tg), gas penetrant permeability and diffusivity, and wide-angle X-ray diffraction measurements. H2- and C2H2-stable AgTf2N-containing membranes could offer an exciting opportunity for increased energy efficiency and process intensification in olefins production via debottlenecking of cryogenic distillation columns with long-lifetime membrane systems.

Abstracts – Oral Presentations, Monday

Presenting authors are indicated by an asterisk*.

Oral 6 – Inorganic Membrane Materials

2:00 pm (Druid, Monday)

Oral 31 - Aramid-derived Carbon Molecular Sieve Membranes for Carbon Dioxide Capture

Gaurav Iyer (University of Maryland)*; Chen Zhang (University of Maryland)

Ten million tons of H2 is produced in the U.S. by steam methane reforming every year with a hundred million tons of CO2 byproduct. Hence, efficient CO2 capture is crucial to sustainable H2 production. Membrane-based H2/CO2 separation can provide energy-efficient alternative to pressure-swing adsorption. Membranes need to provide outstanding H2/CO2 selectivity to be competitive. Carbon molecular sieve (CMS) membranes, made by pyrolysis of polymer precursor membranes, are an emerging class of gas separation membranes with the capability of achieving separation performance exceeding the Robeson upper bound. Several polymer precursors such as polyimides, polybenzimidazoles, and polymers of intrinsic microporosity have been studied as CMS membrane precursors.
In this talk, we will present H2/CO2 separation in ultraselective CMS hollow fiber membranes derived from an aromatic polyamide (aramid) precursor.1 The aramid was synthesized by interfacial polymerization, which decomposed at 400°C with carbon residue of ~50% at 925°C. Aramid-derived CMS membranes showed the highest H2/CO2 ideal selectivity (~360 at 35°C) among all CMS membranes reported in literature. Sorption studies suggested that ultra-high H2/CO2 diffusion selectivity led to the ultra-high membrane selectivity. Aramid-derived CMS membranes also demonstrated outstanding H2/CO2 separation factor of 200 with H2/CO2 mixture feeds at 150°C permeation. These novel aramid-derived CMS membranes will be compared with conventional polyimide-derived CMS membranes to elucidate their ultra-high H2/CO2 selectivity.

Reference:

1. Iyer, G. M.; Zhang, C., Precise Hydrogen Sieving by Carbon Molecular Sieve Membranes Derived from Solution-Processable Aromatic Polyamides. ACS Materials Letters 2023, 5 (1), 243-248.

2:30 pm (Druid, Monday)

Oral 32 - Interfacial Polymerization of Nanoparticles for Modifiable, Ultra-Thin, Ultrafiltration Membranes

Dean F Stipanic (University of Toronto)*

Interfacial polymerization (IPz) is an industrially scalable technique, most notably used in the fabrication of reverse osmosis membranes, with the ability to create highly crosslinked thin films. At an oil/water interface two small molecule monomers react to create extremely thin (< 100-nm), defect free films which are then utilized in a variety of applications. We aim to expand the IPz chemistry library by utilizing 1D nanomaterials, namely SiO2 nanoparticles, as one of the "monomers" creating a highly crosslinked, defect free nanoparticle film for use in membrane separation processes. By controlling the size and surface chemistry of the nanoparticles the properties of the resulting membrane such as surface charge, wettability, and rejection profile can be intentionally tuned. SiO2 nanoparticles are used in this study as a proof-of-concept to validate that nanoparticles can be easily transplanted into the established IPz workflow due to their simple synthesis and modification. However, it can be easily imagined that this approach could be expanded to other nanoparticles and applications such as surface coating, catalysis, or as delivery vehicles.

3:00 pm (Druid, Monday)

Oral 33 - Balanced reduction and Oxidation to Engineer Graphene Oxide Membranes for Dye Desalination

Haiqing Lin (The State University of New York at Buffalo)*; Erda Deng (State University of New York at Buffalo); Kai Chen (University at Buffalo); Mengqi Yuan (University at Buffalo)

Graphene oxide (GO) membranes have been extensively investigated for molecular-sieving membranes for dye removal from textile wastewater because of their tunable channel sizes and ability to permeate water and salts and reject dyes. GOs are often prepared by oxidation of graphene, and they are hydrophilic and can be dispersed in water. Therefore, they must be reduced or cross-linked to improve their stability, which would dramatically lower water permeance. Herein, we demonstrate a balanced reduction and oxidation to manipulate nanostructures and thus dye desalination performance. Specifically, GO nanosheets were first reduced by hydrazine in a solvation state and vacuum-filtrated onto a Nylon substrate and then exposed to peracetic acid (PAA) for in-situ oxidation. The effects of the oxidation conditions (including GO thickness, PAA content, and exposure time) on water permeance and dye rejection are thoroughly investigated. For example, a 100 nm-thick rGO membrane shows water permeance of 35±11 LMH/bar, congo red (CR) rejection of 99.8%, and Na2SO4 rejection, while the PAA oxidation of 10 min increases water permeance by 180% to 93±11 LMH/bar, retain CR rejection of 99.6%, and decrease Na2SO4 rejection to 3.3 %. The oxidized reduced membranes (OrGO) exhibit stable water permeance of 80 LMH/bar and dye rejection over 99.0% in multi-cycle continuous tests in cross flow, superior to state-of-the-art commercial NF membranes and GO membranes reported in the literature. The membranes also display stability in long-term continuous tests with aggressive chemical washing. This work demonstrates an interesting approach to flexibly fine-tune the oxidation and reduction degree of the GO membranes to achieve superior molecular sieving ability.

3:30 pm (Druid, Monday)

Oral 34 - Pyridinic nitrogen substituted two-dimensional pores on single-layer graphene for rapid and selective CO2 transport

Kuang-Jung Hsu (EPFL)*

Single-layer graphene film hosting Å-scale pores has been regarded as an ideal candidate for gas separation due to the fact that atomic thickness has low mass transport resistance allowing high throughput permeance while at the same time yielding an attractive molecular selectivity. However, achieving for similar size gas pair, such as CO2/N2 and O2/N2 by controlling the pore size distribution with a sub-angstrom precision (for high selectivity) and high pore density (for high permeance) has proven challenging. To address this, we incorporate adsorption selectivity in the 2D pore by substituting the edge atom of graphene pores with N (pyridinic N), imparting highly competitive sorption of CO2, leading to a combination of high permeance and high selectivity.
In this presentation, I will introduce single-layer graphene membranes hosting N-substituted nanopores for carbon capture. Briefly, the pyridinic-N-substituted pore is achieved by a facile reaction of oxidized graphene with NH3. The pyridinic-N-substituted pores show a rapid and quantitatively reversible complexation of CO2 with pyridinic N by cycles of adsorption and desorption in the spectroscopy. The phenomenon is also visualized by microscopy where pores are observed occupied and empty upon adsorption and desorption, respectively.
Next, I will discuss the carbon capture performance of graphene hosting pyridinic nitrogen-substituted two-dimensional pores. The highly CO2-philic and N-substituted 2D pores a strong competitive sorption for CO2, which in turn, leads to an attractive performance even in the dilute CO2 source. It led to a large CO2/N2 separation factor (close to 2000) while the atomic selective layer resulted in high CO2 permeance (up to 50000 GPU). Thanks to the uniform and scalable chemistry, high-performance centimeter-scale membranes could be demonstrated. This approach to placing functional heteroatoms at 2D pores opens new directions to push the performance of membranes and sensors.

4:00 pm (Druid, Monday)

Oral 35 - Carbon molecular sieve membranes derived from molecularly engineered cross-linkable 6FDA-based copolyimides for gas separations

Zhongyun Liu (Georgia Institute of Technology)*; William Koros (Georgia Institute of Technology)

Carbon molecular sieve (CMS) membranes offer extraordinary combinations of scalable economical processability and excellent separation performance. In this work, important effects of molecularly engineered cross-linkable 6FDA-based copolyimides are reported. Synergistic benefits of kinked precursor backbones having crosslinkable structures will be shown to exist for both CO2/CH4 and C3H6/C3H8 separations using CMS membranes. Especially when coupled with optimized pyrolysis conditions, CMS materials truly push the state of the art for challenging gas separations to enable decarbonizing key large-scale applications. Beyond dense films, actual asymmetric CMS hollow fibers were further developed, and their performance for CO2/CH4 and C3H6/C3H8 separations will be presented. The work to be discussed spans fundamental and practical features that illustrate why asymmetric hollow fiber polyimide-derived CMS membranes offer optimum scalable solutions to many problems facing the global chemical industry.

4:30 pm (Druid, Monday)

Oral 36 - Ultrasonication Assisted Membrane Surface Functionalization by Silver (Ag)- and Zinc (ZIF-8) Metal-Organic Frameworks for Water Treatment: Synthesis, Characterization, and Performance

Himangshu Mondal (The University of Alabama)*; Medha Kasula (The University of Alabama); Milad Rabbani Esfahani (The University of Alabama)

Metal-organic frameworks (MOF) are well known for their unique properties like high surface area, porosity, tuneability, etc. This precedence makes them highly sought materials of interest in gas storage, catalysis, adsorption, separation, and purification applications. We report the surface functionalization of both Polyethersulfone nanofiltration and thin-film composite (TFC) nanofiltration through in-situ (direct) growth of Ag-MOF and Zn-MOF via ultrasound method. We hypothesize that having such surface functionalization would render facilitated water transport properties, and enhanced selectivity towards specific solutes (contaminant) due to the structure of MOFs. The novelty of the proposed MOF-surface functionalization is the direct growth of MOF using ultrasonication method without damaging the surface of the membrane. The ultrasonication method is facile and environmentally friendly method (compared to the other MOF synthesize that uses high temperature (> 200 °C or harsh solvents) that provides the MOF synthesize within short (as low as 5 mins.) time at ambient conditions. To address the stability of the growing MOF layer concern, the polydopamine was used between the PES and MOFs as a bridging agent. Polydopamine can easily adhere to polymeric substrates and forms coordinate covalent bonding between its catechol (-OH) groups and metal nodes of MOFs, offering strong attachment. The polydopamine (PDA) coated membranes and MOF-PDA membranes were characterized using SEM, XRD, FTIR, contact angle, and zeta-potential techniques. We showed that polydopamine acted as a bridging agent and provided stable MOF-functionalized surface. The MgSO4 salt and Methyl Orange (MO) dye rejection performances of membranes were examined, and 25% and 30% increments were achieved for the MOF-membranes in the salt and dye rejection performances compared to pristine membranes, respectively.

Oral 7 – Industrial Membranes 2

2:00 pm (Yellow Hammer, Monday)
Oral 37 - Development of MTR Polaris Membranes for Carbon Capture

Tim Merkel (Membrane Technology & Research Inc.)*; Johannes G Wijmans (Membrane Technology and Research, Inc.)

For over a decade, Membrane Technology and Research, Inc. (MTR) has been developing a membrane capture technology to decarbonize large point source emitters such as power plants and industrial facilities. This effort has led to the development of the Polaris class of membranes, a novel low-pressure-drop membrane module, and patented process designs.

This paper will provide an update on recent scale-up and testing of our membrane capture process. In 2021/2022, second-generation Polaris membranes packaged in low-pressure-drop planar modules were tested at the Technology Centre Mongstad in Norway. During 6 months of operation treating fluidized catalytic cracker (FCC) offgas, the small pilot system (capacity of up to 20 tonne CO2/day captured) demonstrated CO2 capture rates of >90%. These tests also validated the low pressure drop of containerized planar modules that are the final form factor for commercial deployment of this membrane capture technology.

In current follow-on work, partially funded by the US Department of Energy, a large pilot membrane system is under construction for operation at the Wyoming Integrated Test Center (WITC) in Gillette, Wyoming. This system will capture 150 tonne CO2/day from a flue gas slipstream supplied by the adjacent Dry Fork Station coal-fired power plant. The capture system will produce high-purity liquid CO2 meeting saline injection or CO2 pipeline specifications. When completed later this year, it will be the largest membrane capture plant in the world.

In addition to these technology developments, we will provide an update on industry trends, such as the increasing focus on higher capture rates (deep decarbonization) of hard-to-abate industrial sources, and the potential impact of future membrane improvements.

2:30 pm (Yellow Hammer, Monday)
Oral 38 - Novel Catalyst-Coated Membranes for Green H2 Production

Chunqing Liu (Honeywell UOP)*

Green hydrogen is produced from water and renewable electricity via a water electrolysis process. It is expected to play an important role in the clean energy transition to achieve net zero global emissions by midcentury, particularly in the decarbonization of hard-to-abate sectors, such as gas processing, refining, steel, and petrochemicals. Catalyst-coated membrane (CCM) is a key component for proton-exchange membrane (PEM) water electrolysis and anion-exchange membrane (AEM) water electrolysis for green H2 production. This talk will discuss the breakthrough development of CCMs for green H2 production at Honeywell UOP. The novel CCMs developed at UOP have shown higher electrolyzer efficiency and higher current density than the commercially available CCMs, enabled by the breakthrough proprietary high ionic conductivity membranes, high activity catalysts, and new catalyst coating ink formulas. UOP's CCMs have been tested by a number of leading electrolyzer manufacturers and confirmed to achieve current densities at a factor of 1.3x higher than the commercially available CCMs, enabling about 30% reduction in cost of non-CCM electrolyzer stack components.

3:00 pm (Yellow Hammer, Monday)

Oral 39 - Timeline of Membrane Development

David Hasse (Air Liquide)*; Sandeep Karode (Air Liquide); Marcos Da Conceicao (Air Liquide)

We will discuss the 30 year timeline of the development of a membrane product from the viewpoint of an industrial membrane supplier. From initial discovery of the material in the 1980's, lab work in the 1990's and field trials in the 2000's, this product was commercialized into three markets: hydrogen, biogas and natural gas. From pilot work to successful commercial applications, hydrogen took 2 years, biogas 3 years and natural gas 10 years. The reasons for the differences come down to risk and reward.

3:30 pm (Yellow Hammer, Monday)

Oral 40 - Parametric Analysis of Membrane Properties on Post Combustion Flue Gas System Economics

Joshua C Watson (Compact Membrane Systems, Inc.); Christine Parrish (Compact Membrane Systems)*; Ken Loprete (Compact Membrane Systems)

While the transportation sector can rely on electrification and power generation can utilize renewable sources, the path to industrial decarbonization requires a more innovative approach. Membranes have a unique modular nature and offer simplified systems that are a lynchpins for industrial decarbonization. When applied to post combustion flue gas, membranes can address hard to abate sectors like cement and steel which are responsible for 15% of the global CO2 emissions.

This talk will discuss the system design consideration for post combustion flue gas systems. We will discuss the superstructure modelling approach used to enable membrane staging optimization for given membrane performances to find the lowest overall cost. To understand the tradeoffs, a parametric analysis was performance with membrane properties between 100-5000 GPU and 10-300 selectivity. Due to compression costs, high performance at low pressures is a necessity for economic carbon capture (CC) with membrane systems. In the hard-to-abate CC applications studied, it was found that above a threshold selectivity, increased permeance unlocks significantly lower costs of capture but increased selectivity yields diminishing economic returns. Realistic capture costs were calculated with amortized installed capital estimates and industry relevant utility, maintenance, and labor cost estimates.

Because membranes with increased properties can be more expensive to manufacture, the implications of membrane costing were analyzed to determine the economic benefits of permeance and selectivity. According to these results, a high permeance membrane with modest selectivity (i.e., 1000 GPU and 20 CO2/N2 selectivity) can achieve capture costs under $50/ton with conservative costing assumptions. A high selectivity membrane with modest permeance (i.e., 200 GPU and 100 CO2/N2 selectivity) cannot achieve capture costs under $90/ton with the same costing assumptions.

4:00 pm (Yellow Hammer, Monday)

Oral 41 - Energy Efficient Olefin-Paraffin Separations with Optiperm™ Membrane System

Brandon J Burghardt (Compact Membrane Systems)*; Christine Parrish (Compact Membrane Systems); Ken Loprete (Compact Membrane Systems)

The separation of light olefins from paraffins is one of the most energy intensive in the world. The state of the art separation method is distillation, a technology that has remained unchanged over decades. Membranes hold the promise of a more modular, energy efficient and capital efficient method for separation. However, they have historically been hampered by low performance and stability. CMS has developed a transformational membrane technology that is being demonstrated at commercial scale with Braskem.

The first Optiperm™ membrane utilizes a silver embedded polymeric material to selectively transport olefins. This material has demonstrated over 3 years of longevity in laboratory operation and has shown high selectivity and permeance under realistic operating conditions.

In continuing the commercialization of Optiperm™ membranes, CMS' has partnered with Braskem to jointly develop

and fabricate a multistage pilot unit to test commercial scale spiral wound membranes in in a mixed C3 process stream. This talk will discuss results from the pilot unit that was deployed in a Braskem Polymer facility in January 2022 and was brought on stream in May 2022. This is the first commercial scale operation of Optiperm™ membrane modules under field conditions using actual process streams. We will describe the pilot unit design which incorporates all the process equipment of a commercial scale unit including an advanced process control system that enables the unit to simultaneously produce two high quality product streams.
Learning from our field testing with Braskem will be shared with effects of the process operating conditions and membrane operating window. Updates will also be presented on the scale up of our flat sheet spiral wound commercial membrane cartridges, second generation hollow fiber membrane configuration development, and the progress on our joint development project with Braskem.

4:30 pm (Yellow Hammer, Monday)
Oral 42 - Enabling a Robust Hydrogen Economy via Advanced MEA and Membranes
Noah D Meeks (Southern Company Services, Inc.)*

The hydrogen economy has immense potential to enable the decarbonized energy systems, but advanced technologies including advanced membrane electrode assemblies and membranes can enable the economic feasibility of hydrogen production and utilization. This talk will review the key features of the hydrogen economy, key technical objectives, and discuss novel technical approaches for increasing hydrogen production efficiency, lowering capital costs, and improving safety and performance.

Oral 8 – Catalytic and Responsive Membranes

2:00 pm (Black Warrior, Monday)
Oral 43 - Thermoresponsive antifouling ultrafiltration membranes from mesophase templating
Saadatgharehbagh Younes (University of Oklahoma); Seyed Mostafa Tabatabaei (University of Oklahoma); Kyungtae Kim (Los Alamos National Laboratory); Reza Foudazi (University of Oklahoma)*

Nanostructured polymers synthesized from lyotropic liquid crystal (LLC) templates with normal hexagonal (H1) structure exhibit a 3D-continuous transport path, which makes them ideal for membrane separation application. Incorporating additional functionality, especially stimuli-responsiveness, in such ordered structure provides even more application opportunities. In this work, we present the first successful synthesis of H1-structured thermoresponsive ultrafiltration (UF) membranes via LLC templating. The membrane contains thermoresponsive Poloxamer-diacrylate which not only acts as the monomer and structure-directing amphiphile, but also enables the pore size change with temperature. Experimental studies reveal that the swelling capacity of H1-templated polymer as well as permeability and selectivity of the obtained membrane can be altered via changing the temperature. Increasing the temperature from 25 to 45 °C increases the normalized flux from ~ 28 to ~ 68 L.m^-2.h^-1.µm and molecular weight cut-off (MWCO) from 2200 to 3900 Da. Furthermore, the membrane shows an outstanding fouling resistance against different solutes.

2:30 pm (Black Warrior, Monday)
Oral 44 - Engineering a biocatalytic membrane for caffeine upcycling
Hadi Rouhi (University of Alabama)*; Meredith Mock (University of Alabama); Ryan Summers (The University of Alabama); Steven T Weinman (The University of Alabama)

Xanthines are purine bases found in many foods and drinks. They have different applications in medical settings, such as as central nervous system stimulants, bronchodilators, coronary dilators, diuretics, and anti-cancer adjuvant treatments. All xanthine derivatives have stimulatory effects on the body, but they differ in their potency and duration of action. Caffeine is the most well-known xanthine, and is found in coffee, tea, and some soft drinks. Caffeine (1,3,7-trimethylxanthine) is the most abundant sources of methylxanthines with a low price, while other compounds such as para-xanthine and 7-methylxanthine cost more than 13,000 to 31,000-fold in price. Biocatalytic membranes are membranes that utilize biomolecules on and/or within their porous structures to perform specific catalytic reactions.
In this project, CDE, an iron-based protein, is covalently immobilized on the surface of a polyvinylidene difluoride (PVDF) membrane to impart biocatalytic properties to the membrane. The CDE protein is known to catalyze the reaction of 7-methylxanthine to xanthine (the second step of caffeine upcycling). A redox graft-polymerization was used to modify the base PVDF membrane with an amine-based monomer, methacrylamide. The primary amine

functionalities on the PVDF surface were then modified with glutaraldehyde, which is a known coupling agent for enzyme attachment. After the polymer modification, CDE enzymes were attached to the membrane surface to produce an enzyme-immobilized membrane. Membranes were characterized by SEM, WCA, BET, and ATR-FTIR. The amount of enzyme attached was measured by performing a mass balance. Bradford reagents tests were used for qualitative evidence that enzyme attachment occurred. The activity of CDE enzyme on the membrane surface is measured by evaluating the ability of immobilized enzymes on the membrane to demethylate 7-methylxanthine to xanthine, which is detectable with HPLC.

3:00 pm (Black Warrior, Monday)

Oral 45 - Reverse Osmosis Silica Fouling Control using Reactive Micromixing

Ruizhe Xu (University of Texas at Austin)*; Weiliang Bai (University of Texas at Austin); Manish Kumar (University of Texas at Austin)

Membrane fouling and concentration polarization (CP) are two main issues that increase energy consumption and the cost of the reverse osmosis (RO) membrane separation process. Reactive membranes with hydrogen peroxide (H2O2) has been reported to remove colloidal silica, humic acid (HA), and E. coli foulant[1] in nanofiltration system, as well as oil[2] and bovine serum albumin (BSA) [3] in ultrafiltration system, by generating bubbles and cause micromixing. Dissolved silica has a combined effect with organic acids on membrane fouling [4,5], and reactive membranes have not been applied to prevent this fouling. This study aims to use hydrogen peroxide-assisted reactive membrane for dissolved silica and silica-HA synergetic antifouling in an RO system. The effect of reactive membranes on reducing CP is also reported.

3:30 pm (Black Warrior, Monday)

Oral 46 - Composite Hollow Fiber Membrane Reactor Containing Immobilized Organocatalysts and Palladium for Sustainable Chemical Transformation

Ali A Rownaghi (U.S. Department of Energy/Cleveland State University)*

Catalytically active asymmetric membranes were developed by crosslinking a polydimethylsiloxane (PDMS) thin layer onto a porous polyamide-imide hollow fiber (PAIHF) support, followed by grafting of aminosilane with hydroxyl derived-PDMS/PAIHF, and finally palladium nanoparticles (PdNPs) immobilization using salicylic aldehyde. Aminosilane and salicylic aldehyde linkers were used to permanently immobilize PdNPs onto the PDMS surface through metal coordination chelation, which prevented their agglomeration and leaching from the catalytic membrane reactor (CMR) module. The obtained CMRs were used as a heterogeneous catalyst and continuous-flow membrane reactor for hydrogenation of 4-nitrophenol, aldol and nitroaldol condensation, Heck coupling, CO2 cycloaddition and hydroxyalkylation of aniline, and tandem reactions of glucose and fructose to 5-hydroxymethylfurfural (HMF). Our findings also revealed that the turnover frequency (TOF) and selectivity can be tuned and controlled by adjusting the chemistry and degree of cross-linkers, reaction solvents, and flow rates. Even though our polymeric hollow fiber microreactors showed relatively good performance at temperatures up to 150 °C, some amount of active spices (e.g., Pd nanoparticles) leached out from the microreactor due to polymer swelling, plasticization, and pore shrinkage during flow reaction, especially when exposed to polar aprotic solvents and aromatics, and deteriorated the stability of the immobilized catalysts.

4:00 pm (Black Warrior, Monday)

Oral 47 - Conversion of waste biomass to platform chemicals and fuels using a catalytic membrane reactor

Ranil Wickramasinghe (University of Arkansas)*; Xianghong Qian (University of Arkansas); Chhabilal Regmi (University of Arkansas); Zhexi Zhu (University of Arkansas)

Global warming due to release of greenhouse gases coupled with increasing world energy demands requires immediate development of renewable resources for production of fuels and chemicals. Though many technologies exist to convert biomass feedstocks to fuels and value-added products, commercialization has been slow. The high cost of production of biobased chemicals and fuels frequently makes them uncompetitive compared to fossil fuel derived products. Here a patented catalytic membrane reactor has been developed that can convert a variety of lignocellulosic biomass feed stocks (corn stover, corn fiber, wheat straw, almond and walnut shells, coffee ground, rice husks, food waste etc.) to sugars, hydroxymethylfurfural (HMF) and levulinic acid with high yield and high selectivity. HMF and levulinic acid are considered critical platform chemicals for the production of numerous chemicals, resins, fibers etc as well as intermediates for the production of aviation fuels.
Our catalytic membrane reactor consists of a ceramic membrane substrate immobilized with a dual functional catalyst

by grafting polystyrene sulfonic acid (PSSA) and polyvinyl imidazolium chloride (polyionic liquid, PIL) chains from the surface of the membrane. The PSSA chains catalyze biomass hydrolysis and the subsequent conversion to levulinic acid. The neighboring (PIL) chains help solubilize lignocellulosic biomass and enhance the catalytic activity of the PSSA chains. The PSSA chains were synthesized via surface-initiated atom-transfer radical polymerization whereas the adjacent PIL chains were synthesized via UV-initiated free radical polymerization.
In this presentation the catalytic membrane will be described. Results for production of sugars, HMF and levulinic acid will be presented. The catalytic membrane technology developed here could provide a more economically viable method for utilizing our vast waste biomass thus promoting a circular economy.

4:30 pm (Black Warrior, Monday)
Oral 48 - Effect of PPE decontamination processes on porous polysulfone (PSf) flat sheet membranes fabricated via nonsolvent-induced phase separation (NIPS): A comparison with traditional PPE
Ebuka Ogbuoji (University of Kentucky)*; Isabel Escobar (University of Kentucky)

Air filters are essential for various systems, including HVAC, PPE, microelectronics, and automobiles. The COVID-19 pandemic has emphasized the importance of HEPA filters, leading to an expected 7.1% annual growth rate in the air filter market, projected to reach $22.15 billion by 2028. However, conventional air filters are non-biodegradable and single-use, contributing to the global polymer waste crisis. Membrane technology presents a sustainable solution, achieving high particulate filtration efficiency (>99.5%) through mechanical mechanisms. This study compared the impact of regeneration processes on surgical masks, N95 masks, and polysulfone flat sheet membranes using methods that has been reported to inactivate aerosolized microbes by 100% on commercial mask surfaces. The treated PSf membrane showed no toxic residual elements and retained PEG on the membrane surface. While treatments resulted in a 70% reduction in NaCl aerosol filtration efficiency for masks, there was no significant decrease in PSf membrane filtration performance. Surface potential measurements showed a correlation between static charges and filtration efficiency in commercial masks, which was not observed with the PSf membrane. Mechanical strength tests showed no compromise in membrane strength. These results suggest that flat sheet membranes can be the reusable next generation of air filters for critical applications.

Oral 9 – Electrochemical Separations 1

2:00 pm (Capstone, Monday)
Oral 49 - Advanced electrochemical membranes for capacitive deionization
Christopher G Arges (The Pennsylvania State University)*; Tanmay Kulkarni (The Pennsylvania State University); Bharat Shrimant (The Pennsylvania State University)

Membrane capacitive deionization (MCDI) experienced a renaissance in the past decade due to interest in low energy brackish water desalination. After years of research, there is much debate as to whether MCDI will compete with electrodialysis and reverse osmosis in terms of specific energy use and cost for brackish desalination. The gains in MCDI technology; however, have not been wasted. New materials, cell architectures, and models may position the technology for selective ion separations. In terms of materials research for MCDI, most of the effort has been on electrode materials development (e.g., redox active polymers and other battery-type intercalating electrodes). Less attention has been given to the polymeric ion-exchange membranes (aka electrochemical membranes) for MCDI. This talk will commence with our past research examining how electrochemical membranes' area specific resistance affects brackish water desalination. It will then present our new research on using highly conductive poly(arylene) cation exchange membranes and anion exchange membranes for desalination across the salinity spectrum (from seawater feeds to brackish water feeds). The poly(arylene) ion-exchange membranes consist of all-carbon backbones and are synthesized via superacid catalyzed polymerization route. The talk will close with our recent work on bipolar membrane MCDI for selective ion capture. Bipolar membranes modulate the process stream's pH value. Leveraging solution pH and electrode potential, based upon Pourbaix diagram principles, we are able to achieve selective separations of copper(II) ions from waste streams in one example and recovery of itaconic acid from brine mixtures in another example.

2:30 pm (Capstone, Monday)
Oral 50 - Ion Containing Membranes For Clean Energy and Sustainable Environment

Abhishek Roy (NREL)*

In the coming decade, membrane science will play an important role in addressing non thermal based separation needs for various clean energy technologies and helping the world to achieve a sustainable environment. Specific applications like, H2 utilization and production, carbon capture and reuse, plastic deconstructions, chemical process separations and water purification are noteworthy. On one hand each of these applications has specific performance metrics, on the other hand they share a common fundamental question; relation of membrane structure to transport of various species. In addition to understanding the structure-property relationships, membrane fabrication, and prototype testing are additional research areas for accelerating membrane development. These areas sit at the intersection of engineering, basic science, and science policy and partnerships. There is a need to develop a cross-functional, inter-agency membrane platform initiative to drive such goals. The presentation will focus on key membranes application areas like hydrogen fuel cells, water purification, waste plastic circularity, and carbon capture, emphasizing recent developments, market and societal needs, and proposed research topics.
The presentation will focus on electrochemical separation and how polymer science and critical to achieve improved separation. As an example, fundamental topics like how phase separated morphology influences proton transport in proton exchange membranes will be discussed. The presentation will highlight ion containing block copolymers for proton exchange membranes. Fundamental topics like extent of phase separation and how that influences self-diffusion coefficient of water, proton transport, states of water, methanol permeability will be shared. The presentation will also lead into next generation membrane design guidelines for electrochemical separation.

3:00 pm (Capstone, Monday)

Oral 51 - Bio-Inspired Approaches to Manage Ion Transport Limitation and Directionality of Ion Transport in Electrochemical Systems

Shudipto K Dishari (University of Nebraska-Lincoln)*

The behavior of ionomers in bulk membrane separators and catalyst binder layers of electrochemical devices can be drastically different. Ion and mass transport limitations can make the electrochemical reactions at H-fuel cell cathodes sluggish. This demands new approaches to alter ionomer and interfacial design to deal with geometric and interfacial constraints. This talk will present our recent approaches to manipulate the distribution of ion transport within sub-micron thick ionomeric layers between membrane separators and electrodes. In addition, the talks will share our latest designs of ionomers where the capabilities of biological transport systems are translated, giving a drastic improvement in interfacial ionic conductivity. Some of these ionomers also exhibit voltage-activated and directed ionic current which can bring new advancement in many energy and electrochemical separation processes.

3:30 pm (Capstone, Monday)

Oral 52 - Impact of hierarchical cation-exchange membranes' chemistry and crosslinking level on electrodialysis demineralization performances of a complex food solution

Elodie Khetsomphou (Laval University, Institute of Nutrition and Functional Food (INAF))*; Laurent Bazinet (Laval University, Institute on Nutrition and Functional Food (INAF))

Ion-exchange membranes (IEMs) are integral part of electrodialytic processes that have various applications in food industries including wastewater treatment, whey and fruit juices deacidification, etc., without major impact on the environment. But the high cost of IEMs, due to their complex fabrication methods and highly engineered precursors, significantly contribute to the investment cost of these processes, hindering their applications at industrial scale. Therefore, hierarchical cation-exchange membranes (hCEMs) fabricated by blade coating and UV crosslinking of ionomer on top of a porous substrate have been developed as an alternative. The use of commodity precursor as an ion-exchange layer in combination with energy efficient fabrication processes ensures a low-end cost for hCEMs. These membranes demonstrated promising results performing NaCl demineralization. In the food industry, complex solutions are used and hCEMs were never investigated before for such applications. Two chemistries (urethane acrylate: UL and acrylic acid: EbS) with various crosslinking densities were formulated. The impacts of hCEMs properties and crosslinking density on whey demineralization performances by electrodialysis (ED) were evaluated and compared with CMX, a high performing CEM for whey demineralization by ED. The crosslinking density had an impact on the hCEMs area specific resistance, and on the ionic conductance for EbS membrane. However, 70 % demineralization of 18 % whey solution was reached for the first time for hCEMs without any fouling observed, and with comparable performances to the CMX benchmark. Although some properties were impacted by the crosslinking density, the global performances in ED (limiting current, energy consumption, current efficiency, etc) for EbS and UL6 membranes were similar to the CMX benchmark. These promising results suggest the possible application of hCEMs for whey demineralization by ED, more generally as an alternative in the food industry.

4:00 pm (Capstone, Monday)

Oral 53 - Factors Influencing Ion Transport in Activated Wafer Electrodeionization

Leticia Santos de Souza (University of Arkansas)*

Electrodeionization (EDI) is an ion removal membrane process driven by electrical current widely used for several applications, including water treatment, electronics, pharmaceutical, etc. Cation and anion exchange membranes are strategically positioned between two plate electrodes to create diluted and concentrated chambers so that when electric power is applied, ions are transported from diluted to concentrated compartments. Ion exchange resins (IER) in diluted chambers function as media where the ions flow through, overcoming a process limitation at low concentrations.

Activated wafer electrodeionization (AWEDI) utilizes highly permeable mesh electrodes in inner compartments of EDIs allowing for an independent modulation by electrically turning on or off regions within the device. The initial objective of the mentioned work was to enable multiple combinations of IER with different properties working under the electric current, resulting in different transport rates or selectivity depending on which IER contributes to the removal process.

Nevertheless, the ion transport variations due to electrode positioning in a device with only one type of IER needed to be investigated to clarify the theory behind this innovation. Thus, several AWEDI configurations using the same IER were designed, assembled, and tested. The results suggest that variables such as membrane positioning, concentration gradient, and charge gradient are the factors that guide the different ion transport when the electrode positioning varies.

4:30 pm (Capstone, Monday)

Oral 113 - Ion-exchange biosponge polymers for selective removal of toxic ionic chemotherapy drugs

Hee Jeung Oh (Pennsylvania State University)*

Cancer is becoming the leading cause of death in most developed nations. Despite efforts to develop targeted and personalized cancer therapeutics, dosing of the cancer chemotherapeutics is limited by toxic side effects. Typically, more than 50-90% of the injected drug is not trapped in the target organ and bypasses the tumor to general circulation, causing toxicities in distant locations. Currently we do not have any routes to remove these untrapped toxic chemotherapy drugs other than relying on the body's natural metabolism.

In the context of reducing the toxicity of chemotherapy, we have designed charged, ion-exchange biosponge adsorbers for capturing ionic chemotherapy drugs before they spread through the body. Doxorubicin, a widely used chemotherapy drug with significant toxic side effects, is a model drug. The charge concentration and water uptake in the ion-exchange biosponge adsorbers are systematically changed to achieve high binding capacity and fast kinetics. We aim to understand the mechanism of ionic drug transport in these biosponge adsorbers and design highly selective polymers in the presence of competing species in blood. Using a swine model, our initial design enables the capture of 69 % of the administered drug without any adverse effects. The proposed approach may help patients fight cancer.

Oral 10 – Desalination and Osmotic Processes

2:00 pm (Private Dining, Monday)

Oral 55 - Membrane Design Criteria and Practical Viability of Pressure-driven Distillation

Weifan Liu (Vanderbilt University)*; Shihong Lin (Vanderbilt University)

Pressure-driven distillation (PD) is a novel desalination technology based on hydraulic pressure driving force and vapor transport across a hydrophobic porous membrane. In theory, PD offers near perfect rejection for non-volatile solutes, chlorine resistance, and the ability to decouple water and solute transport. Despite its advantages, pore wetting and the development of a reverse transmembrane temperature difference are potential critical concerns in PD, with the former compromising the salt rejection and the latter reducing or even eliminating the driving force for vapor transport. We herein present an analysis to evaluate the practical viability and membrane design principles of PD with a focus on the dependence of flux and salt rejection on membrane properties. By modeling the mass transfer in a PD process under different conditions, we arrive in two important conclusions. First, a practically detrimental reverse transmembrane temperature difference does not develop in PD under all relevant circumstances and is thus not a practical concern. Second, for a PD process to achieve an acceptable salt rejection, the membrane pores should be at the nanometer scale with a highly uniform pore size distribution. This analysis demonstrates the practical viability of PD and provides the principles for designing robust and high-performance PD membranes.

2:30 pm (Private Dining, Monday)

Oral 56 - Nitrate and Salt Transport Coefficients Determination for Spiral-Wound RO Membranes Derived from Field Data of Treatment of Impaired Brackish Groundwater

Nora Marki (University of California, Los Angeles); Yang Zhou (Shanghai University); Kwan Lam Yip (University of California, Los Angeles); Christian Aguilar (University of California, Los Angeles); Yakubu Jarma (University of California, Los Angeles); Bilal Khan (California State University, San Bernardino); Yoram Cohen (Professor)*

Reverse osmosis (RO) membrane treatment of impaired brackish groundwater often must handle the removal of multiple contaminants. In the present work, salinity reduction and nitrate removal, which is of particular concern in agricultural regions, is analyzed to extract the fundamental membrane solute transport coefficients and water permeability, and assess their dependence on system operating conditions. The above membrane transport parameters were extracted based on high-resolution field data from a spiral-wound RO system (4,000 gallons/day capacity) treating/desalting impaired groundwater of ~50-65 mg/L nitrate as nitrogen, and salinity in the range of ~1,200 – 1,500 mg/L total dissolved solids. The RO treatment system operation was monitored with seventeen process and water quality sensors (e.g., pressure, flow rate, conductivity, nitrate concentration, temperature) whereby system operation was over a period of about 12 months, covering a range of operating conditions (e.g., permeate recovery, feed nitrate concentration, and temperature). System performance was first described via machine learning models to identify the most significant attributes that affect membrane/process performance. The relative importance of each attribute was quantified by its Gini importance, and the incremental contribution to the R2 of the derived model. Solute passage (also quantified in terms of the solute transport coefficient) was most significantly affected by feed temperature and membrane water permeability, also demonstrating a high correlation of nitrate passage with salt passage. The ratios of salt to nitrate passage and salt/nitrate transport coefficients were also evaluated based on models that account for the impact operating conditions, demonstrating the performance of up to R2~0.960. The above results suggest the potential use of correlation of transport parameters for process optimization/control and for assessing nitrate sensor fault/drift and for data imputation.

3:00 pm (Private Dining, Monday)

Oral 57 - Effectively Using Heat to Thermally Enhance Osmotic Processes for Power Generation and Desalination

Sanjana Chintalacheruvu (Oakland University)*; Saber Khanmohammadi (Oakland University); Jonathan Maisonneuve (Oakland University)

Thermal energy from sources such as solar radiation and industrial waste heat is abundant and represents a tremendous resource for improving the sustainability of water and energy infrastructure in the US and throughout the world. In this presentation, we examine the potential of using low-grade heat to improve the performance of both power generation by pressure retarded osmosis (PRO) and of seawater desalination by reverse osmosis (RO). A combination of laboratory testing and numerical simulations are used to evaluate thermally-enhanced processes at temperatures between 20–50 °C. Results indicate that power from thermally-enhanced PRO can be more than doubled and that the specific energy of thermally-enhanced RO can be reduced by as much as 15 %. An improved finite element model is validated and used to identify transport dynamics that account for improvements and tradeoffs. For example, it is found that heating feed in PRO is preferable to heating draw, because thermal polarization is relatively limited, and therefore hot feed produces a warmer and less viscous permeate, resulting in an apparent increase in permeability. A new metric of effectiveness is defined to compare increases in PRO power and RO energy relative to thermal energy input. In general, operation at ~ 30 °C is found to be most effective, but best operating temperatures are also shown to be highly case-specific and dependent on a number of key variables including the value of thermal energy relative to electric energy. This analysis, therefore, provides insight into the potential for using heat to improve the performance of osmotic processes for a variety of energy and water applications.

3:30 pm (Private Dining, Monday)

Oral 58 - Effect of Dehydration on RO Desalination Performance in Fully Aromatic Polyamide Membranes Containing TEMPO-Oxidized Cellulose Nanocrystals (TOCNs)

Connor L Farrell (Virginia Tech)*; E. Johan Foster (University of British Columbia); Stephen M Martin (Virginia Tech)

Lack of access to safe drinking water is an increasing problem across the world. One of the most effective methods for producing drinking water is the desalination of seawater or brackish water. Fully aromatic polyamide thin film composite (TFC) membranes are the most commonly used material for desalination by reverse osmosis. Thin film nanocomposite membranes (TFNs) incorporate nanoparticles in the selective layer of TFCs to improve their

performance. Both TFCs and TFNs are limited by a tradeoff between flux and salt rejection and by limitations in use conditions. Of particular interest for this study is the fact that drying of TFC membranes is known to decrease their performance, likely due to a collapse of the pore structure of the polyamide resulting in decreased water transport. This potentially limits the ease with which RO membranes can be stored and transported, as they must remain hydrated to maintain performance, or must undergo a potentially time-consuming rehydration process. We have created novel TFNs by incorporating functionalized rod-like cellulose nanocrystals (CNCs) in the polyamide layer resulting in improved desalination performance, particularly when membranes are dried and rehydrated.
The CNCs used for this study were modified to produce 2,2,6,6-Tetramethylpiperidine-1-oxyl (TEMPO)-oxidized cellulose nanocrystals (TOCNs). CNCs are selected for their high aspect ratio, low cost, availability, and sustainability. The flux and salt rejection were investigated for TFNs containing varying amounts of TOCNs. TFNs were either stored in ultrapure water or dried under vacuum prior to testing to investigate effects of dehydration. Water flux increased with the addition of TOCNs without any decrease in sodium ion rejection, and the maximum water flux was observed at an intermediate loading level. There was a 22% maximum increase in water flux relative to the unloaded hydrated membrane and 146% maximum increase relative to the unloaded dried membrane.

4:00 pm (Private Dining, Monday)

Oral 59 - Reverse osmosis and membrane distillation: potential tools for leveraging surplus energy at nuclear power plants for seawater desalination and societal benefit

J. Hunter Himes (Tennessee Tech University); Laura H Arias Chavez (Tennessee Tech University)*

Nuclear power plants represent a significant, untapped supply of low-grade waste heat and intermittently available surplus electricity that could be used for seawater desalination. The majority of thermal energy generated at these facilities is discharged to the environment as low-grade heat; only one third is converted into electricity. In addition, nuclear power production in the US cannot be ramped up or down to match fluctuating demand for electricity, forcing generation facilities to sell electricity at a loss at times of low grid demand. With a suitable partner process, this surplus energy could be used to desalinate seawater without significantly increasing energy demand; this also improves the economics of nuclear power generation. A range of desalination technologies with potential to leverage surplus electricity or waste heat were investigated. Preliminary screening based on thermodynamics and technology readiness identified intermittently operated reverse osmosis (RO) and membrane distillation (MD) as top candidates. Process-level material and energy balances were used to model intermittent RO, MD, and a RO-MD hybrid process in MATLAB. The quantity and quality of water that could be produced from given amounts and types of energy under varying operating conditions were predicted. Models were applied to a hypothetical case study in which desalination processes were built on-site and simultaneously with a new 300 MW nuclear power plant in Turkey Point, Florida with a projected 50-year lifespan. The viability of each possible scenario was assessed according to the price that would have to be charged for the desalinated water to break even on costs over the lifetime of the facility. These results inform efforts to utilize waste heat and surplus electricity in membrane processes, and they illuminate the potential benefits that can accrue for both the local community and the generator of the surplus energy.

4:30 pm (Private Dining, Monday)

Oral 60 - Pressure-induced diffusion of solvents in dense homogeneous membranes

Kevin Reimund (University of Texas at Austin)*; Rahul Sujanani (University of California, Santa Barbara); Jorge Hernandez (University of Texas at Austin); Kristofer Gleason (University of Texas at Austin); Manish Kumar (University of Texas at Austin); Benny Freeman (The University of Texas at Austin)

Knowledge of the nature of solvent transport through polymers is crucial for developing future membranes for application in solvent separations. Classically, two main theories have been proposed to describe the pressure-driven flux of solvent across a membrane: "pore-flow" models which assumes the permeating solvent conducts mechanically through percolating pores and "solution-diffusion" models which treat the membrane as an independent liquid phase through which the permeating solvent partitions and diffuses. A key feature of the solution-diffusion model is the existence of a concentration gradient during permeation which was demonstrated in a seminal work by Rosenbaum and Cotton. Paul and Ebra-Lima later demonstrated the existence of such concentration gradients via similar technique for the pressure-driven permeation of organic solvents through crosslinked natural rubber.

Recently, we have used similar techniques to demonstrate the existence of pressure-induced concentration gradients in water-swollen materials such as Nafion 117 and poly(styrene sulfonate) with a high-pressure reverse osmosis sweep cell capable of accommodating stacks of membranes up to ~4 mm in total thickness. We now extend this technique to investigate the permeation and flux of organic solvent and water through these materials, cellulose acetate films of various degrees of acetylation, and crosslinked poly(ethylene glycol diacrylate), at pressures up to 3000 psi. The existence of pressure-induced concentration gradients of solvent either precludes, or may demand modification of, pore-based explanation of solvent flux over diffusion-based ones, particularly as it relates to designing novel solvent-separating membranes.

Abstracts – Oral Presentations, Tuesday

Presenting authors are indicated by an asterisk*.

Oral 15 – Emerging Materials for Liquid Separations

9:30 am (Private Dining, Tuesday)
Oral 61 - Solvent-Free Carrier-Based Membranes for Selective Ion Separations

Jay Werber (University of Toronto)*

Conventional membrane materials that separate based on size or charge typically exhibit poor selectivity for cations of similar valency, which is needed in important applications such as battery recycling, e-waste recycling, resource recovery from wastewater, and mineral production. Carrier-based membranes incorporate dissolved, freely diffusing ligands, which form selective complexes with target ions and "carry" the ions through the membrane. These ligands are typically commercial hydrophobic extractants used in hydrometallurgical solvent-extraction processes. While carrier-based membranes have enabled extraordinary selectivity, the typical formats considered---supported liquid membranes and emulsion liquid membranes---are too unstable for industrial use. Polymer inclusion membranes---also called "gelled supported liquid membranes"---use a plasticized glassy polymer matrix and can offer moderately enhanced stability, but still ultimately become unstable and are typically impossible to regenerate.

This talk will detail our lab's ongoing work with solvent-free carrier-based membranes, with the goal of creating ion-selective membranes that are stable enough for industrial use. Our approach is to use a rubbery polymer layer as a "solvent" in which the carriers can dissolve without added plasticizers. I will focus on our recently published study (DOI: 10.1021/acs.est.2c07242) in which we made polymeric microcapsules as solid, solvent-free mimics of emulsion liquid membranes. For this iteration, we used commercial polystyrene-b-polybutadiene-b-polystyrene triblock polymer as the capsule walls, with the copper-selective ligand Lix 84-I diffusing through the rubbery polybutadiene phase. Dilute sulfuric acid formed inner aqueous cavities. The capsules rapidly took up copper with negligible sodium uptake, and showed 15-fold selectivity for Cu(II) over Ni(II). Packed beds of the capsules enabled complete Cu(II) removal in 2 minutes, with reusability to at least 10 cycles.

10:00 am (Private Dining, Tuesday)
Oral 62 - Structure and transport studies of carbon molecular sieve membranes for wastewater treatment

Young Hee Yoon (Georgia Institute of Technology)*; Ryan Lively (Georgia Tech)

The demand for effective water resource utilization is rising due to the growing global population, economic development, and the adverse effects of climate change. Wastewater reclamation is critical in addressing this demand. However, a considerable amount of wastewater is still being discharged without proper treatment. In particular, industrial wastewater, which accounts for 16 percent of global freshwater withdrawals, is an important area to focus on.

The removal of dissolved small molecule organic solvents, such as BTEX, from water, is a key challenge in industrial wastewater treatment. Carbon molecular sieve (CMS) materials offer a promising solution for scalable, high-performance reverse osmosis (RO) membranes. Their chemical and thermal stability, and the ability to precisely alter the pore size distribution, offer a significant advantage in optimizing membrane performance for a wide range of separation applications.

We hypothesize that the rigid and carbonaceous CMS membranes can effectively separate dissolved organic solvents from water by providing enthalpic and entropic separation contributions compared to flexible membranes. Previous work has shown that p-xylene selective separation can be achieved through vapor permeation and that CMS membranes can potentially remove p-xylene from water with high energy efficiency in pervaporation (PV) systems.

This study investigates the performance of hollow fiber CMS membranes in pressure-driven RO and activity-driven PV systems. We explored the liquid separation properties of CMS membranes fabricated under different pyrolysis conditions to control the transport in the membranes. Our results show that all CMS membranes exhibit molecular sieving separation in RO, while organic-selective separation is hypothesized in PV. Our research aims to lay the foundation for utilizing CMS hollow fiber membranes for effective water-organic liquid separation and optimizing the CMS microstructure to control the separation.

10:30 am (Private Dining, Tuesday)

Oral 63 - Differences in water and vapor transport through angstrom-scale pores in atomically thin membranes

Piran Kidambi (Vanderbilt University)*

The transport of water through nanoscale capillaries/pores plays a prominent role in biology, ionic/molecular separations, water treatment and protective applications. However, the mechanisms of water and vapor transport through nanoscale confinements remain to be fully understood. Angstrom-scale pores (~2.8–6.6 Å) introduced into the atomically thin graphene lattice represent ideal model systems to probe water transport at the molecular-length scale with short pores (aspect ratio ~1–1.9) i.e., pore diameters approach the pore length (~3.4 Å) at the theoretical limit of material thickness. Here, we report on orders of magnitude differences (~80×) between transport of water vapor (~44.2–52.4 g m−2 day−1 Pa−1) and liquid water (0.6–2 g m−2 day−1 Pa−1) through nanopores (~2.8–6.6 Å in diameter) in monolayer graphene and rationalize this difference via a flow resistance model in which liquid water permeation occurs near the continuum regime whereas water vapor transport occurs in the free molecular flow regime. We demonstrate centimeter-scale atomically thin graphene membranes with up to an order of magnitude higher water vapor transport rate (~5.4–6.1 × 104 g m−2 day−1) than most commercially available ultra-breathable protective materials while effectively blocking even sub-nanometer (>0.66 nm) model ions/molecules.

References

1. Cheng et al. Nature Communications 2022
2. Cheng et al. ACS Applied Materials and Interfaces 2022
3. Cheng et al. Nano Letters 2020

11:00 am (Private Dining, Tuesday)

Oral 64 - Removal of Micro- and Nano-plastics from Water using Liquid-Liquid Extraction

Ashish Srivastava (The University of Alabama)*; Milad Rabbani Esfahani (The University of Alabama); Jason Bara (University of Alabama)

Micro- and Nano-plastics (MNPs) are the potential emerging plastic-related pollutants that have adverse impacts spreading across all the environmental media, including fresh water. The current conventional water treatment methods, such as coagulation, flocculation, filtration, photocatalytic degradation, and microbial bioremediation, showed a lack of efficiency in removing different MNPs. Therefore, there is a need for the development of a new removal/extraction process. The liquid-liquid extraction (LLE) method is a promising separation technique for efficiently removing MNPs in water treatment. In the present work, we report the removal of two types of polystyrene (PS) from water MNPs, including synthesized micron-size (>1μm) PS and commercial sulfonated PS (nano-size, ~70 nm) using a hydrophobic IL [C8mim] [Tf2N]. The extraction process showed more than 99% removal of both PS MNPs at the ratio of 1:40 (vol: vol) IL/water for the removal of 100 ppm PS MNPs from water. In addition, the extraction efficiency of PS MNPS was evaluated in the presence of 2000 ppm NaCl, and ~ 99.36 % removal was achieved, showing the selectivity of [C8mim] [Tf2N] towards PS MNPs.

11:30 am (Private Dining, Tuesday)

Oral 65 - Synthesis and fabrication of mesoporous membranes with enhanced mechanical properties from amphiphilic multiblock polymers via self-assembly and nonsolvent-induced phase separation

Ben Pedretti (University of Texas at Austin)*

Isoporous membranes that combine high pore density with narrow pore size distributions are some of the most promising candidates for the next generation of ultrafiltration membranes because they are not limited by the same permeability-selectivity tradeoff more traditional ultrafiltration membranes exhibit. Although significant advances have been made since the discovery of self-assembly and nonsolvent-induced phase separation (SNIPS) as a method for producing highly selective membranes, significant work remains to make these membranes an industrially viable technology. One of the largest drawbacks of the polymers that have shown great promise with SNIPS processing is their lack of mechanical integrity. To address this critical issue, mechanically robust, mesoporous membranes were cast from solutions of novel ABAC tetrablock polymers polystyrene-b-poly(ethylene-alt-propylene)-b-polystyrene-b-poly(ethylene oxide) (SESO) and polystyrene-b-polysoprene-b-polystyrene-b-poly(4-vinylpyridine) (SISV). The polymers were synthesized via sequential anionic polymerization to molecular weights between 65 and 110 kg/mol. Small angle X-ray scattering (SAXS) was used to verify polymer morphology in the casting solutions, while scanning electron microscopy (SEM) was used to verify the structure of the resultant membranes. The rubbery midblock between the two polystyrene domains that make up the structural matrix of the membrane substantially increased the toughness and flexibility of the membranes over membranes fabricated from polymers that use polystyrene as the only structural block. This added toughness allowed for the casting freestanding membranes with strain energy densities two orders of magnitude higher than a PS-P4VP membrane of significantly higher molecular weight. Our membranes also exhibited pure water permeances up to ca. 2100 LMH/bar, an order of magnitude higher than a track etched membrane with double the pore diameter, with pore sizes as low as 23 nm.

12:00 pm (Private Dining, Tuesday)
Oral 66 - None

Oral 12 – Downstream Bioprocessing

9:30 am (Yellow Hammer, Tuesday)
Oral 67 - Characterization and Associated Pressure-Dependent Behavior of Deposited mRNA-Lipid Nanoparticles Formed During Sterile Filtration
Kevork Oliver Messerian (The Pennsylvania State University)*

The recent success of mRNA vaccines has generated increased interest in the use of Lipid Nanoparticles (LNP) for delivery of nucleic acids. However, the large size and unique behavior of the LNP can create significant challenges in the design and implementation of the manufacturing process, including bioburden reduction and sterile filtration steps. Recent studies have demonstrated that low capacity and high degree of fouling during LNP filtration at constant pressure is due to pore blockage by deposited LNP. The objective of this study was to characterize the mRNA-LNP deposit formed on the surface of sterile filtration membranes using a combination of electron microscopy imaging, atomic force microscopy, and hydraulic resistance measurements. Analysis of the microscopic characterization data and hydraulic resistance data will be discussed for mRNA-LNP deposits formed on a series of microporous membranes with different pore size. Taken together, these results provide important insights into the physical properties of the fouling deposit and its impact on the sterile filtration of LNP used in the production of these important biotherapeutic products.

10:00 am (Yellow Hammer, Tuesday)
Oral 68 - Fouling of Virus Clearance Membranes
Yair Kaufman (TeraPore Technologies)*; Rachel Dorin (Terapore); Gabriel Hale (TeraPore Technologies); KC Hunt (TeraPore Technologies); Daniele Gerion (TeraPore Technologies)

Virus filtration—often called virus clearance—is the last downstream process of manufacturing of monoclonal antibodies (mAbs) for therapeutic applications. One of the main challenges during virus filtration is membrane fouling, which is often attributed to the existence of mAb aggregates that block the membrane pores. Therefore, to mitigate pore-blocking fouling, an intense effort is directed in the development of appropriate prefilters that selectively block aggregates and allow the passage of non-aggregated mAbs.
In this presentation, we will shed light on fouling that is governed by mAb adsorption to the membranes, which is rarely discussed in the field of virus filtration. Adsorptive fouling cannot be mitigated by a prefilter; instead, it calls for the development of new surface chemistries that reduce the adhesion force/energy of mAbs to the membrane, and thereby reduce mAb adsorption.
The results that we will present show that, in the case of isoprene-styrene-4-vinylpyridine membranes (TeraPore Technologies), mAb adsorption is governed by hydrophobic forces, and less by the double-layer forces (i.e., electrostatic, and entropic forces). Furthermore, we will describe a filtration system and a simple theoretical model that together allow for quantifying mAb adsorption, as well as distinguishing between pore-blocking from adsorptive fouling mechanism.

10:30 am (Yellow Hammer, Tuesday)
Oral 69 - Surrogate-based Optimization for the Techno-economic Feasibility Analysis of Membrane Capture Chromatography Platforms
Juan J Romero (Clemson University)*; Scott Husson (Clemson University); Eleanor Jenkins (Clemson University)

Recently, we developed a computational framework to simulate and optimize the capture chromatography process for monoclonal antibody purification, which was used to compare the performance of membrane- and resin-based platforms. The process simulation incorporated dynamic models for affinity chromatography that were validated with experimental breakthrough curves. The results were integrated with an Intelligen SuperPro Designer process simulation to evaluate key performance indicators of the operations. The resulting simulation was robust, adaptable, and capable of providing the information needed for decision-making in specific production contexts. Nevertheless, the use of multiple simulation blocks (including approximating nonlinear partial differential equations with finite elements to generate breakthrough curves) made this framework computationally expensive and difficult to optimize. To reduce the computational time and make the framework more attractive for industrial applications, we created a library of breakthrough curves to generate surrogate functions used by the optimization algorithm. This strategy yielded accurate results with a 92% decrease in processing time. Additionally, the surrogate function can use experimental or simulated breakthrough data, an important advantage in situations where the available models cannot represent the system, experimental data can be easily obtained, or there are no resources for the simulation of dynamic systems. In this work, we present the newly developed framework and used to optimize cost of goods and process time for the case study of an industrial-scale membrane-based capture chromatography platform.

11:00 am (Yellow Hammer, Tuesday)

Oral 70 - Valorization of Food Byproducts inside Processing Industries

Shanti Bhushan (GEA)*

Food processing industries play an important role in food security by processing and preserving foods. The food byproducts generated during the process become factory effluent costing money to treat. In addition, the byproducts also become environmental burden. The role of the presentation is to discuss case studies of value addition to diverse food byproducts generated in different food processing industries.

The use of plant proteins is increasing for its perceived benefits to health. About 10-25% of proteins are lost as wastewater during the processing. Introduction of membranes filtration after proper chemical and physical pretreatments helps recover the protein from the effluent as well as minimizes the overall energy consumed in the process.

Production of meat by cell culturing process requires rejuvenation and reuse of the cell culturing growth media for economical scalability. The growth media consists of amino acids, growth factors, buffer, sugar etc. Membrane filtration economically purifies the used media and make it ready for recycling back in the process.

Polyphenols found in several fruits, vegetables, tea, and coffee provide multiple health benefits due to their antioxidant properties. During the production of olive oil, the byproduct olive mill wastewater (OMWW) containing abundance of phenolic compounds is disposed of. By using membrane filtration, polyphenols from OMWW is extracted and purified. The recovered polyphenols are added to food or used in cosmetics for boosting filtering of sun's ultraviolet rays.

Collagen is a major structural protein found in animal bones, tendons, skin and other connective tissues. It is a byproduct of animal meat industry. Gelatin as well as collagen peptide are made using the collagen. Membrane filtration plays a constructive role in limiting damage of collagen peptides, reducing microbial degradation during processing and reducing cost of production.

11:30 am (Yellow Hammer, Tuesday)

Oral 71 - Virus adhesion to archetypal membrane surfaces: Developing guidance on membrane selection for virus removal

Vlad Tarabara (Michigan State)*; Xunhao Wang (Michigan State University)

The presentation will describe recent studies aimed at elucidating physicochemical bases of virus adhesion to common membrane surfaces. Adhesion of human viruses (enveloped and non-enveloped) and bacteriophages onto several "archetypal" surfaces is predicted using the extended Derjaguin-Landau-Verwey-Overbeek theory and experimentally assessed in tests with quartz crystal microbalance with dissipation. The results are interpreted based on physicochemical properties of viruses and membranes. The surfaces are typified on the basis of their surface energy components including the dispersive (Lifshitz-van der Waals) component and two polar (electron donor and electron acceptor) components. Surfaces include common polymers, polyelectrolyte multilayers, ceramics, and passivated metals. We show that hydrophobic and electrostatic interactions govern virus attachment while van der Waals interactions play a relatively minor role. In higher ionic strength solutions, the extent of virus attachment correlates with the free energy of virus-surface interfacial interaction. The proposed methodology can guide screening and selection of membrane materials with controlled virus adhesion. The information on the efficiency of virus attachment to membranes as a function of their surface energy components can help design membrane materials, develop membrane cleaning solutions and protocols, and inform transport and fate models for viruses in water treatment systems.

References

[1] Wang, X.; Tarabara, V. V. Virus adhesion to archetypal fomites: A study with human adenovirus and human respiratory syncytial virus. Chem. Eng. J. 429 (2022) 132085

[2] Lasareishvili, B.; Shi, H.; Wang, X.; Hillstead, K. D.; Tediashvili, M.; Jaiani, E.; Tarabara, V. V. Virus recovery by tangential flow filtration: A model to guide the design of a sample concentration process. (Note.) Biotechnol. Prog. 31 (2021) e3080.

12:00 pm (Yellow Hammer, Tuesday)
Oral 72 - Improved Filtration Performance of Virus Filters at High Product Titers

Rony Horax (University of Arkansas); Wenbo XU (University of Arkansas); Xianghong Qian (University of Arkansas); Ranil Wickramasinghe (University of Arkansas)*

Recent improvements in cell lines, media, and bioreactor performance have led to large increases in product titer, creating significant challenges for the downstream purification process. This includes the virus filtration step, which is a critical component of the overall virus clearance strategy in the production of monoclonal antibodies (mAbs) and plasma-derived products. There is also growing interest in the possibility of performing virus filtration as part of the final fill-finish operations (in combination with the sterile filtration), which would provide added insurance against virus contamination while eliminating a separate stand-alone virus filtration step.

However, the capacity and fouling behavior of virus filtration at such high product titers (30 g/L or above) can be compromised by the formation of reversible and possibly irreversible product aggregates as well as by increased product-membrane interactions. In addition, increased viscosity or concentration polarization at high product titers could also reduce the filtration flux.

Up till now, the exact fouling mechanism that leads to significant flux decay was not understood. Here mAb feed streams at high product titers were characterized and correlated with virus filtration performances. The methods used include size exclusion chromatography, SDS PAGE and UV spectroscopic measurements. In addition, mAb solubility at different buffer conditions was determined. It was found that reversible product aggregates formed at high product titers are responsible largely for the flux decay during virus filtration.

Oral 13 – Membrane Synthesis and Casting 2

9:30 am (Black Warrior, Tuesday)
Oral 73 - Towards better understanding of synthesis-morphology-performance in polyamide membranes: Visualization under controlled reaction conditions

Adi Ben Zvi (Technion)*; Guy Z Ramon (Technion - Israel Institute of Technology)

Thin-film polyamide (PA) films fabricated via interfacial polymerization (IP) form the core of membrane-based water purification. Despite widespread commercial application, the link between synthesis conditions and resultant film properties is not fully understood. Unravelling this fundamental knowledge is essential for designing better membranes for different applications. Much experimental work has been reported, where synthesis conditions are varied, and the resultant performance and morphology are discussed. However, contradicting trends are often observed, primarily due to the lack of rational isolation of physical parameters affecting the reaction as well as the use of post-fabrication characterization techniques. Here, we study synthesis-morphology-performance relationship in PA membranes by systematically varying reaction parameters so as to isolate their effect on physical properties of the system. The reaction dynamics are observed and analyzed using in-situ visualization via microscopy-based particle tracking; fluorescent particles located in the aqueous phase are observed by confocal microscopy during an IP reaction performed in a microfluidic device. Particle trajectories and motion parameters are calculated, determining the aqueous-phase flow nature as the indication of hydrodynamic instabilities that occur during IP. The underlying assumption here is that IP can induce an interfacial instability that alters film morphology. A transition from stable to unstable state corresponds to a morphological shift of the resultant film, from smooth to crumpled. Resultant film morphologies are also characterized using SEM, AFM and TEM images as well as performance parameters of the fabricated membranes. This presented methodology allows us to close the loop in the fundamental understanding of synthesis-morphology-performance relationship in PA membranes, which is an important step towards the transition to alternative green materials for membrane fabrication.

10:00 am (Black Warrior, Tuesday)
Oral 74 - Molecularly mixed composite membranes based on triptycene-isatin Porous Polymer Networks and Carboxylic-acid Functionalized PIM-1

Lucas Condes (University of Oklahoma)*; Matthew T Webb (University of Oklahoma); Adriana Landry (University of Oklahoma); Michele Galizia (University of Oklahoma)

Porous polymer networks (PPNs) are hyper-crosslinked 3-dimensional networks that have shown promise as gas sorbents and as fillers in mixed-matrix membranes (MMMs) for gas and liquid separations. When PPNs are incorporated in glassy polymers to form MMMs, however, a variety of behaviors are observed, which can lead to either an increase or decrease of permeability. A decrease in selectivity with respect to the PPN-free polymer is usually observed.
To promote an intimate polymer-PPN contact and finely tune the microstructure, porosity, and selectivity of PPN MMMs, we incorporated triptycene-isatin PPN into carboxylic-acid functionalized PIM-1 (PIM-COOH). PIM-COOH is found to undergo a radical-induced crosslinking reaction as low as 200 °C, and convincing evidence is given from SEM, TGA, NMR, and gel fraction analysis that this crosslinking occurs not only among PIM-COOH chains, but also between PIM-COOH and PPN. To systematically study the effect of polymer-PPN mutual crosslinking on the structural and transport properties of the resulting MMMs, four different membranes were studied, namely neat PIM-COOH, neat PIM-COOH with thermal treatment, PIM-COOH-PPN MMMs without thermal treatment, and PIM-COOH-PPN MMMs with thermal treatment at 200°C. The effect of this low-temperature crosslinking reaction, as well as the presence of PPNs, on the gas transport properties and plasticization characteristics of the resulting MMMs was studied by structural characterization via NMR, microscopy, free volume calculations, and independent gas permeability and solubility measurements and application of the dual mode sorption-partial immobilization model. By doing so, we were able to deconvolute the roles of the PPN in thermally treated and untreated membranes with and without inclusion of the PPN. This study opens new routes to synthesize high performance separation systems for targeted applications by simply tuning the polymer and filler chemistry and interactions.

10:30 am (Black Warrior, Tuesday)
Oral 75 - Functionalized NF and adsorptive membranes for the separation of perfluoro-organics and environmentally relevant ions

Francisco Leniz (university of Kentucky)*; Dibakar Bhattacharyya (U. Kentucky)

The current complexity of water streams and the need for separating both value as well as undesirable solutes from water keeps demanding higher efficiency from separation platforms. Adequately matching the properties of the solute and the separation technologies can enable less energy-intensive, selective, and scalable processes. Membranes have been a reliable and versatile technology to treat water. Modifying current commercial membranes extend the application portfolio for membranes, with opportunities for obtaining better affinities and separation yields. Thus, helping solve critical separation needs while creating less complex downstream scenarios and lowering energy requirements.
This work introduces new functionalized membranes for the application on separations of organics from water, such as PFAS, and the recovery of inorganic ions such as lanthanides. Different hydrogels with strong and weak acid and basic acrylate functional groups were incorporated into membranes with 30 nm pore size. Adsorption studies were performed against commercial ion exchangers. The strength of the functional groups was the dominant parameter. Both hardness and PFAS removal were tested to compare the separation efficiencies, and the favorable convective flow kinetics were highlighted. The integration of nanofiltration membranes casted on top of commercial 13 nm UF membranes or pore functionalized MF porous support were also investigated to analyze the potential for either enhancing selectivity or to help solve downstream process concerns. Models, including DSPM-DE were used to help understand the potential separation performance of these membranes. Supported by electron microscopy imaging and x-ray photoelectron spectroscopy we identified the localized areas where the functionalization occurred. Finally, the scalability for the functionalized adsorptive membranes was studied and we highlighted some limitations and opportunities for these separation platforms.

11:00 am (Black Warrior, Tuesday)
Oral 76 - Lignin-based membrane fabricated with deep eutectic solvent

Abaynesh Gebreyohannes (King Abdullah University of Science and Technology (KAUST))*; Suzana Nunes (King Abdullah University of Science and Technology (KAUST))

Membrane technology has been driver for sustainable industrial growth through energy efficient molecular separation and waste minimization. Yet, membrane fabrications heavily depend on fossil-based polymers and organic solvents. Identifying bio-derived materials and green solvents is important.
Lignin is a byproduct of the pulp industry at 70 million ton per year, which could be considered for membranes if appropriate fabrication processes with green solvents would be developed. We propose the membrane fabrication using Kraft (alkali) lignin fully dissolved in a green deep eutectic solvent (DES). By far, DES has been integrated into membrane fabrication as supported liquid membranes, an additive or co-solvent for membrane casting in combination with other solvents. This work is the first report of (i) DES used as a single solvent for polymer solution casting in a membrane fabrication, (ii) dissolution of lignin in DES to form a homogeneous solution of polymer concentration 25

wt%, (iii) preparation of pure lignin-based membranes by dissolution and simple casting.
The crosslinking in aqueous medium enhanced the membrane chemical and mechanical stability enabling its application for filtration (molecular weight cut-off 1.3 kg mol-1) in aqueous and organic solvents with perspective of exploration for different applications.

11:30 am (Black Warrior, Tuesday)
Oral 77 - Scaling-Up and Modelling the Fabrication of a More Sustainable Polymeric Membrane for Water Treatment Using Slot Die Coating

David Lu (University of Kentucky)*; Kwangjun Jung (Georgia Institute of Technology); Tequila Harris (Georgia Institute of Technology); Isabel Escobar (University of Kentucky)

Due to increasing clean water insecurity, polymeric membranes are playing a major role as more efficient and scalable filters. A variety of techniques for polymeric membrane fabrication has been studied, with nonsolvent-induced phase separation (NIPS) being one of the most prevalent methods. However, the frequent use of hazardous solvents that are derived from nonrenewable resources hinders the safety and sustainability of NIPS; moreover, new regulations have motivated commercial manufacturers to identify alternative solvents. While bench-scale studies on green solvent-based polymeric membranes have emerged, the next step of manufacturing scale-up remains largely unstudied.
In this study, the next step is addressed by developing a novel scaled-up manufacturing NIPS method for green solvent-based polysulfone (PSf) membranes via slot die coating (SDC). The SDC setup was configured for continuous processing on a roll-to-roll system. PSf dope solutions included Rhodiasolv® PolarClean and gamma-valerolactone (GVL), two nontoxic and biodegradable solvents that have been relatively understudied in membrane synthesis. Dope solution properties were analyzed to determine the optimal parameters for fabricating membranes via SDC. A comparison study of filtration characteristics was conducted between PSf-PolarClean-GVL membranes fabricated via bench-scale blade-casting and SDC. A computational model was developed to determine membrane characteristics based on a large set of SDC parameters.l
Under dead-end filtration conditions, PSf-PolarClean-GVL membranes fabricated via SDC exhibited an average permeability of 55.9 ± 3.7 LMH/bar and a maximum bovine serum album (BSA) rejection of 97.0 ± 5.2%, which indicated close agreement with a blade-casted counterpart (69.7 ± 9.7 LMH/bar and 99.2 ± 1.3%, respectively). The complete results highlight the value of green solvents and SDC in bringing sustainable and scaled-up polymeric membranes closer to implementation.

12:00 pm (Black Warrior, Tuesday)
Oral 78 - Patterning Super-oleophobic Poly (vinylidene fluoride) Membranes via Confined Thermally Induced Phase Separation for Enhanced Oil-Water Separation

Shouhong Fan (University of Colorado at Boulder)*; Duong T. Nguyen (University of Colorado Boulder); Kieran R Fung (Membrane Science, Engineering, and Technology (MAST) Center, Department of Mechanical Engineering, University of Colorado Boulder); Anthony Straub (University of Colorado Boulder); Yifu Ding (Membrane Science, Engineering, and Technology (MAST) Center, Department of Mechanical Engineering, University of Colorado Boulder)

Industrial wastewater containing finely dispersed oil droplets is challenging to treat. Membrane-based oil-water separation technology is lower in cost for the rejection of oil droplets having small sizes down to a few microns. However, the oil-water separation efficiency of the membrane is hindered by oil-fouling. Although chemical modification of membrane surface i.e., through grafting nanoparticles is used to alleviate the oil fouling propensity, the grafted components are susceptible to degradation over long-term operation. Patterns on membrane surface have been experimentally demonstrated for fouling mitigation and improving local hydrodynamic adjacent to membrane surface during the liquid-based separation process. Ultimately, the presence of surface patterns has a great potential to further reduce the extent of membrane oil fouling. To date, the effectiveness of surface patterns on a membrane's oil-water separation performance has not yet been systematically studied.
In this presentation, we demonstrate that our novel patterning technique namely lithographically templated thermally induced phase separation (lt-TIPS) faithfully imparts patterns on membrane surface, therefore, improving oil rejection and maintaining a higher water permeability of membrane. In combination with zwitterionic polymerization-based surface coating, a series of super-oleophobic robust micro-patterned PVDF membranes are fabricated with a controlled coating layer thickness (50-100 nm). The material formulation of lt-TIPS patterned membranes and the mechanism that facilitates conformal coating on patterned surfaces will be discussed. The effect of surface coating on patterning fidelity and pore size will be investigated. In addition, the surface patterns on critical flux and resistance against oil-fouling of membranes will be systematically studied by comparing membrane performance in terms of

oil-water separation and water permeability between patterned and non-patterned membranes.

Oral 14 – Highly Selective Separations

9:30 am (Capstone, Tuesday)

Oral 79 - Specific Ion Selectivity in Sulfonated Polystyrene Membranes Near the Percolation Threshold

Yuxuan Huang (Columbia University)*; Marshall Tekell (Columbia University); Hanqing Fan (Columbia University); Jingchao Qin (Columbia University); Sanat Kumar (Columbia University); Ngai Yin Yip (Columbia University)

Ion-exchange membranes (IEMs) have been successfully deployed in various applications for water, energy, and the environment. Beyond the primary function of charge selectivity, i.e., between oppositely charged counter- and co-ions, there are increasing demands for the discrimination between the like-charged counterions with the same valence, i.e., specific ion selectivity between, for example, alkali metal cations. However, fundamental structure-property understanding of such selectivity in IEMs is still critically lacking. In this study, we leverage differences in the hydrated radii and hydration free energies, -ΔG_hyd, of monovalent cations to engineer cation exchange membranes with tunable specific ion selectivity. Membranes were fabricated by neutralizing polystyrene-r-sulfonated polystyrene (PS-r-SPS) copolymers having controlled sulfonation levels with different cations (Li+, Na+, and K+) and then fabricated into thin films through solvent evaporation. Wide-angle X-ray scattering analysis and ion exchange capacity and conductivity measurements indicate a critical ion cluster percolation threshold for ions to permeate through the membrane. The percolation threshold can be rationalized by the water volume fraction in hydrated membranes. Near the percolation threshold, the K/Li transport selectivity characterized in electrodialysis increases with the decreasing water per charge site in the membrane matrix, λ. Ion sorption results further reveal that selective partitioning is the main factor driving the preferred transport of K+ over Li+ in the fabricated membranes, with the sorption selectivity rationally controlled through λ. The study advances the mechanistic understanding of the specific ion selectivity of IEMs and can inform the rational development of more selective ion separation membranes.

10:00 am (Capstone, Tuesday)

Oral 80 - Chelating DGA membranes enable ion-ion selectivity in radiopharmaceutical processing

Megan Sibley (Case Western Reserve University); Tuli Banik (Case Western Reserve University); Pavithra H. A. Kankanamalage (Brookhaven National Laboratory); Maura Sepesy (Case Western Reserve University); Alexa Ford (Case Western Reserve University); Tim Yen (Case Western Reserve University); Jasmine Hatcher-Lamarre (Brookhaven National Laboratory); Cathy Cutler (Brookhaven National Laboratory); Christine E Duval (Case Western Reserve University)*

Ac-225 is a promising isotope for targeted alpha therapy (TAT, a personalized cancer treatment) that has shown excellent patient outcomes in clinical trials. In the US, clinical demand for Ac-225 far outpaces its availability. One bottleneck in the Ac-225 supply chain is the slow, column-based separation process that operates at a flowrate of 1 mL/min. While resin-based extractive chromatography has long been the workhorse for medical isotope production—these purification schemes are not easily scaled to support clinically relevant production. The purification of Ac-225 requires ion-selective separations to remove Th, Ra, and lanthanides; thus rapid, highly-selective separation materials are required.

To fill this need, we synthesized chelating membrane adsorbers—a scalable, high-throughput alternative to extractive resins. This work spans membrane synthesis (electrospinning), diglycolamide (DGA) ligand synthesis, covalent functionalization of the electrospun membranes, and adsorption/desorption experiments with Ac-225 and relevant competitor ions. The resulting membrane adsorbers contained 36wt% DGA which is consistent with commercially available extractive resins containing physisorbed DGA ligands. Importantly, the DGA-functionalized membranes maintain a high permeance of ~ 500 LMH/bar. Adsorption experiments conducted in 4M nitric acid with La and Ac-225 (both trivalent ions) indicate that the adsorbed Ac-225 is roughly one third of the value predicted by the La isotherm. Rapid desorption kinetics with 10M nitric acid suggest that these DGA-functionalized membranes have the potential to accelerate the Ac-225 purification process. While these chelating membranes are being studied in the context of nuclear medicine, they serve as a promising platform for other applications which require ion-ion selectivity such as rare earth element recovery and radioanalytical chemistry.

10:30 am (Capstone, Tuesday)

Oral 81 - Selective extraction of medium-chain fatty acid from organic waste streams using supported liquid membranes

Hiroki Fukuda (University of British Columbia); Jongho Lee (University of British Columbia)*

Medium-chain fatty acids (MCFAs) produced from organic waste are promising sources for biofuels, pharmaceuticals and cosmetics, but their selective recovery remains a challenge. We develop a supported liquid membrane (SLM) for selective MCFA recovery from organic waste streams. The SLM was fabricated by filling a nonpolar solution of trioctylphosphine oxide (TOPO) in dodecane into a porous and hydrophobic membrane. A feed solution containing ethanol, short-chain fatty acids (SCFAs) and MCFAs was prepared to mimic the effluent of MCFA-producing bioreactors. Placed between the feed solution and a base draw solution, the SLM transfers MCFAs to the draw solution driven by selective absorption into the membrane liquid and the concentration gradient of protonated MCFAs. We demonstrated >90% recovery of MCFAs while ethanol and SCFAs were effectively rejected. The results show that the partition coefficients of different FAs into the membrane liquid modulate the selectivity and recovery. Specifically, a higher concentration of TOPO resulted in a higher MCFA recovery due to the enhanced complexation of MCFAs by TOPO. Conversely, the selectivity of MCFAs over SCFAs increased with a lower TOPO concentration owing to the larger contrast in partition coefficients of SCFAs and MCFAs. Finally, interfacial polymerization technique, which results in a thin barrier film on the SLM surface, was employed to minimize the membrane liquid loss. The enhanced stability of the SLM was also assessed by electrochemical impedance spectroscopy. Our work proves the potential of SLMs for selective MCFA recovery in building a new biochemical production stream from organic wastes.

11:00 am (Capstone, Tuesday)

Oral 82 - Foundational Framework for Evaluating Performance of Nanofiltration-based Solute-solute Separation

Shihong Lin (Vanderbilt University)*; Ruoyu Wang (vanderbilt university)

Membrane filtration has been widely adopted in various water treatment applications, but its use in selective solute separation for resource extraction and recovery is an emerging research area. When a membrane process is applied for solute–solute separation to extract solutes as the product, the performance metrics and process optimization strategies should differ from a membrane process for water production because the separation goals are fundamentally different. Here we used lithium (Li) magnesium (Mg) separation as a representative solute–solute separation to illustrate the deficiency of existing performance evaluation framework developed for water–solute separation using nanofiltration (NF). We performed coupon- and module-scale analyses of mass transfer to elucidate how membrane properties and operating conditions affect the performance of Li/Mg separation in NF. Notably, we identified an important operational trade-off between Li/Mg selectivity and Li recovery, which is critical for process optimization. We also established a new framework for evaluating membrane performance based on the success criteria of Li purity and recovery and further extended this framework to separation with the target ions in the brine. This analysis lays the theoretical foundation for performance evaluation and process optimization for NF-based selective solute separation.

11:30 am (Capstone, Tuesday)

Oral 83 - A Novel Membrane Solvent Extraction Process Enabling Highly Efficient Separation and Recovery of Critical Materials from End-of-Life Lithium-Ion Batteries

Shailesh Singh Dangwal (Oak Ridge National Lab)*; Syed Islam (Oak Ridge National Laboratory); Priyesh Wagh (Oak Ridge National Lab); John Klaehn (Idaho National Laboratory); Ramesh Bhave (Ornl)

In recent years, lithium-ion batteries (LIBs) have drawn a huge attention worldwide due to their wide use in portable electronics and hybrid/electric vehicles. Critical elements, such as lithium (Li), cobalt (Co), and nickel (Ni) are an integral part of these batteries. Although the demand for LIBs is growing rapidly, the production of these critical elements is limited in mining. Separation, recovery, and recycling of these critical elements from end-of-life (EOL) batteries can help in meeting the growing demand. In this work, we report the separation and recovery of critical elements from EOL LIBs using an economically viable and environmentally friendly multistage membrane solvent extraction (MSX) process. In the MSX process, an organic phase consisting of an extractant, and an organic solvent is immobilized into the pores of polymer hollow fiber support. MSX is a continuous modular process. In first stage of MSX process, critical elements, including Li, Co, and Ni are separated from non-critical elements including iron, copper, zinc and aluminum from an acidic feed solution using a cationic extractant called Di-(2-ethylhexyl)phosphoric acid (DEHPA). Subsequently, cationic extractant, Cyanex 272 is used to first separate Co and then mixture of Li and Ni at different pH. Using MSX system, critical elements were recovered with >90% recovery and with 99.9- 99.99 wt% purity from a wide range of spent LIB feedstocks through a series of separation stages. MSX system for each stage also exhibited long-term (~100 h) stability in terms of product purity and recovery. The integrated process requires low chemical inventory and generates minimal waste compared to alternative technologies such as traditional solvent extraction, hydrometallurgy and pyrometallurgy, thereby promoting circular economy. This novel MSX process holds a huge potential for industrial applications because it is easily scalable, energy-efficient, cost-effective, and environmentally friendly.

12:00 pm (Capstone, Tuesday)
Oral 84 - Isolation of Rare Earth Ions Using Electrodialysis or Nanofiltration

Merlin Bruening (University of Notre Dame)*; Dong Ding (University of Notre Dame); Jun Wang (University of Notre Dame)

Rare earth salts are vital in magnets and catalysts for a range of applications. This work examines the possibility of isolating highly charged rare earth ions using either electrodialysis or nanofiltration followed by diafiltration. Partitioning of rare earth ions into ion-exchange membranes is highly selective, but electrodialysis shows only modest selectivities. Low transport selectivities in electrodialysis result from a low lanthanide mobility along with concentration polarization that leads to relatively high diffusive fluxes of the monovalent ion. In contrast nanofiltration shows much higher rejections of lanthanides than sodium. Moreover, concentration polarization can help to decrease the sodium rejection. Selective enrichments of lanthanides in nanofiltration followed by diafiltration leads to highly pure lanthanides.

Oral 11 – Polymeric and Mixed Materials for Gas Separations

9:30 am (Druid, Tuesday)
Oral 85 - Long term stability of thin film cross-linked ionic liquid membranes for CO_2 gas separation

Richard D. Noble (University of Colorado, Boulder)*; Douglas Gin (Department of Chemistry, University of Colorado - Boulder); Chamaal Karunaweera (University of Colorado, Boulder); Matthew Cowan (University of Canterbury, NZ)

CO2/light gas (CH4 and N2) separation performance, physical aging and plasticization of thick film (> 100 microns) cross-linked poly(ionic liquid)-ionic liquid (PIL-IL) composite membranes were previously evaluated at elevated temperatures (up to 50 C) and feed pressures (up to 40 atm). PIL-IL composite membranes made with IL-based multifunctional cross-linkers exhibited higher CO2 permeability and CO2/light gas selectivity compared to the membranes made with a conventional, nonionic cross-linker, divinylbenzene (DVB). For example, a crosslinked PIL-IL composite membrane prepared using 70 wt % 1-vinyl-3-methylimidazolium bistriflimide as the IL monomer, 30 wt % 1-ethyl-3-methylimidazolium bistriflimide as the free IL, and 8 wt % 1,3,5-tris(1'-methylene-3'- vinylimidazolium bistriflimide)benzene as the IL cross-linker exhibited 90% retention of its initial CO2 permeability after testing at 50 °C and 40 atm pressure for 42 weeks (7100 h). In comparison, an analogous PIL-IL composite membrane cross-linked with DVB shattered after testing for 5 h at the same testing conditions. These cross-linked PIL-IL composite membranes can also be formed into thin film composite (TFC) membranes.
A thin film (~5 micron) composite membrane was prepared to make a direct comparison of aging effects with conventional polymer membranes. The membrane was tested periodically for CO2 permeation and CO2/N2 selectivity at 5 bar feed pressure and 20 C. The membrane retained it's initial permeance over 6 months demonstrating that ionic liquid membranes can greatly reduce or eliminate aging effects in polymer membranes used for gas separations.

10:00 am (Druid, Tuesday)
Oral 86 – Rational design of mixed-matrix metal-organic framework membranes for molecular separations

Shuvo Jit Datta (KAUST)*

Conventional separation technologies to separate valuable commodities are energy intensive, consuming 15% of worldwide energy. Mixed-matrix membranes, combining processable polymers and selective adsorbents, offer the potential to deploy adsorbent distinct separation properties into processable matrix. We reported a concept for the construction of a mixed-matrix MOF (MMMOF) membrane based on three interlocked criteria: (i) a MOF filler with optimal pore size and shape, functionality, and a host-guest interaction that selectively enhances H2S and CO2 diffusion while excluding CH4; (ii) tailoring MOF crystal morphology along 001 crystallographic direction into high-aspect-ratio (001) nanosheets that proffer maximum exposure of 1D channel and promote nanosheet-polymer interaction resulting in high nanosheet loading; and (iii) in-plane alignment of (001) nanosheets in a polymer matrix with proximal distance to translate the molecular separation properties of single MOF nanosheets into a uniformly [001]-oriented macroscopic MMMOF membrane. The methodology described here offers great potential for the development of various practical membranes to address numerous industrial energy-intensive separations (1).

[1] S. J. Datta, A. Mayoral, N. M. S. Bettahalli, P. M. Bhatt, M. Karunakaran, I. D. Carja, D. Fan, P. G. M. Mileo, R. Semino, G. Maurin, O. Terasaki, M. Eddaoudi, Science 376, 1080-1087 (2022).

10:30 am (Druid, Tuesday)

Oral 87 - Highly selective polymer membranes containing poly(amic-acid) Stabilized Silver Nanoparticles

Matthew T Webb (University of Oklahoma)*; Lucas Condes (University of Oklahoma); Michele Galizia (University of Oklahoma); Sepideh Razavi (University of Oklahoma)

Facilitated Transport Membranes (FTMs) have been demonstrated to be effective in simultaneously increasing permeability and selectivity, making them a viable approach to overcome the perm-selectivity trade-off experienced by conventional polymer membranes. Despite high performance, metal ion carriers experience photo/chemical aging, which causes, in turn, a rapid decay of their performance. Moreover, when embedded in polymer materials to form mixed matrix membranes, structural defects may form at the metal-polymer interface that, despite accelerating small molecule transport, cause a parallel selectivity loss.

To address these issues, in this study, we fabricated silver nanoparticles (NPs) stabilized with a polyether derived from 4,4-oxydiphthalic anhydride and a polyetheramine. In conjunction with the metal core, polymer agents may be used to not only provide stability against aggregation to the NP, but also tune gas solubility selectivity. Additionally, these polymer agents can mediate an interphase between silver and the bulk polymer, while offering carrier sites protection from photo/chemical degradation.

In this study, nanoparticles for CO2 separation were rationally designed and synthesized by chelating silver nanoparticles with poly(amic-acid)s (PAAs). The effect of the PAA length and ether functional group concentration on the structure and transport properties of nanoparticles were systematically investigated. The silver-PAA nanoparticles were then embedded into a commercial polymer, Pebax, to fabricate defect-free mixed matrix membranes for CO2/CH4 separation, whose structure and transport properties were examined. Remarkably, inclusion of only 1% wt nanoparticles in Pebax enhances CO2 permeability by about 50% and CO2 selectivity by about 100% relative to neat Pebax, while showing excellent signs of stability. The individual contributions of solubility and diffusivity coefficients to the observed permeability behavior were studied in detail.

11:00 am (Druid, Tuesday)

Oral 88 - Tailoring Ultramicropores in Hybrid Carbon Molecular Sieve Membranes for H_2 Purification

Leiqing Hu (University at Buffalo)*; Won Il Lee (Stony Brook University); Soumyabrata Roy (Rice University); Ashwanth Subramanian (Stony Brook University); Kim Kisslinger (Brookhaven National Laboratory); Lingxiang Zhu (National Energy Technology Laboratory); Vinh Bui (University at Buffalo); Shouhong Fan (University of Colorado); Thien N Tran (University at Buffalo, The State University of New York); Gengyi Zhang (The State University of New York at Buffalo); Yifu Ding (Membrane Science, Engineering, and Technology (MAST) Center, Department of Mechanical Engineering, University of Colorado Boulder); Pulickel M. Ajayan (Rice University); Chang-Yong Nam (Brookhaven National Laboratory); Haiqing Lin (The State University of New York at Buffalo)

Porous carbons with tunable pore structures have attracted substantial attention for membrane separation. Particularly, carbon molecular sieve (CMS) membranes with micropores (7 to 20 Å) that promote gas transport and ultramicropores (<7 Å) that render strong size-sieving ability can be prepared from the pyrolysis of polymeric precursors and show excellent gas separation properties, such as blue H2 production from fossil fuel gasification. However, creating suitable sub-4 Å ultramicropores and high microporosity in CMS membranes to sieve H2 from other gases (including CO2, N2, and CH4) is challenging. Here, we report highly H2-selective hybrid CMS membranes derived from a mixed matrix material (sMMM) containing crystalline ZIF-8 and thermally labile Zn/imidazole coordination in polybenzimidazole (PBI). The coordination preserves a tight packing and generates single Zn atoms bonded on polymer-derived carbons after the carbonization, and ZIF-8-derived carbons prevent carbon densification and thus achieve high porosity. Therefore, the obtained CMS membranes have small sub-nano pores of 3 – 4 Å offering strong size-sieving ability and high porosities up to 33% promoting gas transport, leading to both increased H2 permeability and selectivity. More importantly, those sub-nano pores can be tuned to achieve precise cut-offs of H2 against specific gases. The 3.64 Å cut-off forms after Tc reaches 550 °C. And when Tc is 700 °C, H2 permeability reaches to 430 Barrer with H2/N2 and H2/CH4 selectivity at 310 and 1000 at 100 °C, respectively. Particularly, The CMS membrane carbonized at 900 °C exhibited H2/CO2 selectivity of 130 at 70 °C, higher than those of most leading membranes. Moreover, the membrane showed stable and superior H2/CO2 separation performance (H2/CO2 selectivity of ≈60 and H2 permeability of ≈55 Barrer at 100 °C) under simulated gas conditions in 120 h. This study opens a new approach to designing advanced CMS materials for H2 separation and other applications.

11:30 am (Druid, Tuesday)

Oral 89 - Analysis of gas transport in molecularly-mixed composite membranes

Matthew Rivera (MIT)*; Ryan Lively (Georgia Tech)

Individual molecules with intrinsic porosity, such as porous organic cages (POCs), have significant potential to improve the performance of a variety of separations media. An exemplar application is the blending of POCs with polymers to make molecularly-mixed composite membranes (MMCMs). The intimate interaction between individual cage molecules and polymer chains results in a "solid-solution" that avoids longstanding interfacial issues associated with mixed-matrix membranes. Moreover, as the cages are soluble in polymer solutions, the processing of these composites can be easily adapted to established polymer-based technologies as concerns with two-phase processing systems are avoided. MMCMs are still a relatively new development, and underlying transport processes within the membrane are not well understood. Here, we offer a detailed interpretation of guest transport through these solid solutions. We demonstrate how the presence of cage molecules affects polymer chain motions that can impact guest transport through the polymer phase. We also show how cage loading affects membrane free volume. We find that gas permeation deviates significantly from predictions made with the Maxwell model for mixed-matrix membranes. POCs were found to significantly alter membrane properties in the polymer phase because of intimate molecular interactions between the POC and polymer, violating one of the Maxwell models underlying assumptions. This work provides preliminary information on the nature of guest transport in MMCMs to aid their future adaptation to industrially-relevant separation units.

12:00 pm (Druid, Tuesday)

Oral 90 - Interfacial poly(epoxyether) membranes, a versatile thin-film synthesis platform for carbon capture membranes

Daan Van Havere (KU Leuven)*; Rhea Verbeke (KU Leuven); Benny Freeman (The University of Texas at Austin); Ivo Vankelecom (KU Leuven)

Ethylene oxide (EO)-based gas separation (GS) membranes are one of the leading candidates for membrane-based post-combustion carbon capture, due to their favorable combination of CO_2/N_2 selectivity and permeability. Amongst them, crosslinked PEO (XLPEO) materials have shown some of the highest permeability and selectivity combinations in literature. However, processing XLPEO membranes into thin films through conventional techniques can be challenging due to their crosslinked nature.

This works aims to circumvent this issue through an interfacial design strategy, based on the ring opening polymerization of epoxides, resulting in poly(epoxy ether) (PEE) thin-film composite GS membranes. This synthesis procedure presents three major advantages: (1) the interfacial polymerization (IP) synthesis intrinsically yields thin films, (2) the chemistry of PEE films results in crosslinked membranes with a high content of CO_2-philic ether bonds, and (3) the flexibility of the epoxide chemistry enables the incorporation of a variety of monomers and functional groups into the membrane.

The influence of multiple reaction parameters such as, reaction time, densification steps and monomer and initiator concentration and ratio, on the IP synthesis was studied for a model system based on poly(ethylene glycol) diglycidyl ether as the epoxide monomer. Next to influencing membrane properties such as film morphology and integrity, XPS measurements confirmed that varying the monomer/initiator ratio during synthesis influences the membrane charge density as well as its performance. Permeation measurements showed high CO_2/N_2 mixed gas selectivities up to 58, but low permeances due to pore infiltration of the PEE film in the membrane support. The pore impregnation issue was mediated through changing the monomer, significantly increasing the membrane permeance. This work explores to the best of our knowledge for the first time the high potential PEE IP platform for development of GS membranes.

Abstracts – Oral Presentations, Tuesday

Presenting authors are indicated by an asterisk*.

Oral 16 – Molecular and Process Modeling

2:00 pm (Druid, Tuesday)

Oral 91 - Molecular Dynamics Simulations of Charge Interactions in Polyamide Membranes

Suwei Liu (Northwestern University); Sinan Keten (Northwestern University); Richard Lueptow (Northwestern University)*

When implementing nanofiltration (NF) membranes for water treatment and contaminant removal, the charged functional group concentration in the membrane, which depends on the feed pH, has an effect on flux and solute rejection. While size exclusion plays an important role in solute rejection, membrane charge interactions with solutes can affect the selectivity of small mono-/divalent ions for polyamide membranes. To better understand solute-membrane charge interactions, molecular dynamics simulations are used to model the filtration processes. Previous atomistic models have assumed membrane neutrality and only focus on single solute type in the feed. In this study, we account for the effect of feed pH on membrane charge and show how the amount and location of charge in the membrane affects ion selectivity and rejection. We also bridge the gap of membrane experiments and previous computational studies by incorporating multiple solutes in the feed to uncover solute-solute and solute-membrane charge interactions. This work offers insights into the influence of membrane charge and membrane structure on solute selectivity, which can be used to engineer membranes with desired features. Funded by NSF Grant CBET-1840816.

2:30 pm (Druid, Tuesday)

Oral 92 - Water Model Choice for Molecular Dynamics Simulations of Polyamide Membranes

Suwei Liu (Northwestern University)*; Sinan Keten (Northwestern University); Richard Lueptow (Northwestern University)

Polyamide membranes are widely implemented for water treatment and contaminate removal. To better understand the interactions between membrane, water, and solutes, molecular dynamics simulations are often used to model the filtration processes. While previous efforts focused on building realistic membrane models to study the membrane filtration process, the atomistic water model and simulation parameters also can influence calculations. Previous polyamide membrane simulation studies used a variety of water models, so the dependence of water model effect on water transport remains unclear. In this research, we investigate three water models, SPC/E, TIP3P, and TIP4P, and find that they have significant impact on water flux and diffusivity in simulations of a charged nanofiltration membrane system. We also compare the water dynamics results from molecular dynamics simulations with previous experimental and computational results. In general, TIP4P is the most appropriate water model to represent realistic water dynamics in polyamide membrane filtration systems. Funded by NSF Grant CBET-1840816.

3:00 pm (Druid, Tuesday)

Oral 93 - Physical mechanism of water transport in reverse osmosis membranes: Solution-diffusion or pore-flow?

Li Wang (Tongji University)*; Mohammad Heiranian (Yale University); Menachem Elimelech (Yale University)

We performed non-equilibrium molecular dynamics (NEMD) simulations and solvent permeation experiments to unravel the mechanism of water transport in reverse osmosis (RO) membranes. The NEMD simulations reveal that water transport is driven by a pressure gradient within the membranes, not by a water concentration gradient, in marked contrast to the classic solution-diffusion model. We further show that water molecules travel as clusters through a network of pores that are transiently connected. We then conducted permeation experiments with water and organic solvents using polyamide and cellulose triacetate RO membranes, demonstrating that that solvent permeance depends on the membrane pore size, kinetic diameter of solvent molecules, and solvent viscosity. This observation is not consistent with the solution-diffusion model, where permeance depends on the solvent solubility. Motivated by these observations, we then demonstrate that the solution-friction model, in which transport is driven by a pressure gradient, can describe water and solvent transport in RO membranes.

3:30 pm (Druid, Tuesday)

Oral 94 - Machine Learning Modeling of Intermittent Reverse Osmosis Desalination System Operation via Transfer Learning and Long Short-term Memory

Yang Zhou (Shanghai University); Nora Marki (University of California, Los Angeles); Christian Aguilar (University of California, Los Angeles); Kwan Lam Yip (University of California, Los Angeles); Yakubu Jarma (University of California, Los Angeles); Bilal Khan (California State University, San Bernardino); Yoram Cohen (Professor)*

Reverse osmosis (RO) membranes design and operational modes are not standardized. Therefore, operational differences are common among field-deployed RO systems; hence, the interest in developing a modeling approach whereby a given system model can be leveraged to accelerate model development for a similar system. Accordingly, in the present work, we integrate transfer learning (TL), deep learning (DL) Artificial Neural Networks (ANNs), and a similarity-based long short-term memory (S-LSTM), to describe both the unsteady-state and steady-state operational modes of different (and geographically separated) field RO systems treating impaired groundwater. In the modeling process, Transfer AdaBoost (TrAdaboost) approach was combined with S-LSTM to utilize the learned features (i.e., weights of links within S-LSTM model) trained based on 196 days of RO system operation according to the pairwise similarity among model attributes. The learned features were reused to construct a new S-LSTM model of the same structural configuration to describe the new RO system operation based on only a few weeks of its operational data. The approach resulted in significant improvements in forecasting accuracy of system operation owing to the implemented TL strategies combined with S-LSTM. The forecasting performance of the model for a new system, despite measurable differences between the different RO systems with respect to system design (i.e., capacity, hydraulic components, membrane elements) and operating conditions, was with R2 of 0.88 and average absolute relative error (AARE) of 5.7% compared with a lower model performance (R2 of 0.61 and AARE of 19.3%) without transfer learning. The current study suggests the potential use of transfer learning for RO system performance optimization and fault detection, along with domain knowledge from multiple existing systems of similar configurations to alleviate constraints imposed by limitations of data availability.

4:00 pm (Druid, Tuesday)

Oral 95 - Impact of dynamical correlations on salt transport in solvated ion exchange membranes

Nico Marioni (The University of Texas at Austin)*; Akhila Rajesh (The University of Texas at Austin); Zidan Zhang (The University of Texas at Austin); Harnoor Sachar (The University of Texas at Austin); Benny Freeman (The University of Texas at Austin); Venkat Ganesan (The University of Texas at Austin)

Ion exchange membranes are widely used in water purification and energy storage applications to selectively and efficiently regulate salt transport, yet the influence of dynamical ion-ion correlations and ion pairing/condensation on salt transport in charged polymer membranes remain poorly understood. In this study, we use the framework of Onsager transport coefficients within atomistic molecular dynamics simulations to study the impact of ion-ion correlated motion on salt transport in experimentally relevant hydrated polymer membranes. At sufficiently high salt concentrations, cation-anion dynamical correlations exert a significant influence on both salt diffusivities and conductivities. Anion-anion distinct correlations, arising from the imbalance between the concentration of free (mobile) cations and anions, and the retarding effect of the fixed charge groups on cations, proves to be an additional important feature for charged membranes. Our results demonstrate that dynamical correlations should become an important consideration in experimental measurements of salt diffusivities and conductivities for non-dilute salt solutions in polymer membranes.

4:30 pm (Druid, Tuesday)

Oral 96 - Monovalent Ion Selectivity of Cellulose Acetate Membranes

Paul R Irving (University of Texas at Austin)*; Kevin Reimund (University of Texas at Austin); Everett S Zofchak (The University of Texas at Austin); Nico Marioni (The University of Texas at Austin); Benny Freeman (The University of Texas at Austin); Venkat Ganesan (The University of Texas at Austin)

Increasing demand for lithium requires improved methods for extraction from brine. Magnesium contamination is associated with low yield in traditional production processes. Polymer membranes offer a potential alternative for selectively enriching lithium. Using dense cellulose acetate membranes, our experiments indicate a remarkably high lithium/magnesium permeability selectivity on the order of 870:1. The reported selectivities of asymmetric reverse osmosis, nanofiltration, and selective electrodialysis membranes are typically less than 50. Further, our experiments demonstrate that the high selectivity of cellulose acetate is attributed to differences in both salt diffusivity and solubility. We produce atomistic molecular dynamics simulations that support the observed selectivity trends. Higher charge of magnesium relative to lithium is found to result in a strengthening of ion-water interactions relative to ion-polymer interactions, resulting in reduced sorption and hindering the hopping of ions between water clusters. These observations are used to explain the observed experimental trends.

Oral 17 – Organic Solvent Separations

2:00 pm (Yellow Hammer, Tuesday)

Oral 97 - Transport Property Modulation via Non-Aqueous Solvent Specific Behavior in Cross-Linked Membranes

Charles Leroux (University of Virginia)*; Patrick McCormack (University of Virginia); Geoffrey M Geise (University of Virginia); Gary Koenig (University of Virginia)

The push towards a sustainable world has driven vast improvement in renewable energy technologies, but current energy storage systems limit their deployment due to inherent intermittency. Non-aqueous redox flow batteries (RFBs) provide a promising energy storage option due to economic advantages associated with scale-up and design flexibility. However, the enabling technologies for these systems still need further development and understanding. To advance the technology towards real world application, a membrane with high ionic conductivity and low permeability to the dissolved species which store/deliver energy (redox shuttles) must be developed. Membrane properties are influenced by the specific solvent used in the application because of solvent-specific effects, such as solvent uptake. Understanding the influence of the solvent on critical membrane properties is important for directing engineering of non-aqueous redox flow battery membrane technology.
Here, we report a study aimed at decoupling the effects of non-aqueous solvent on the solvent uptake and transport properties of cross-linked polymer membranes. Specifically, a poly (phenylene oxide) (PPO) polymer was functionalized with phenoxyanaline tri-sulfonate (POATS) fixed charge groups and cross-linked. We investigated the influence of three different solvents (acetonitrile, propylene carbonate, and dimethyl carbonate) on ionic conductivity and redox molecule permeability, as all three solvents are of interest for RFB applications. We determined that differences in polymer solvation between each of these solvents resulted in solvent specific ionic conductivity and redox molecule permeability properties. Our results suggest that selecting the appropriate solvent (or appropriately matching the membrane / solvent system) could lead to simultaneously increased conductivity and reduced redox molecule permeability, and such an increase in the selectivity is favorable for enhancing flow battery efficiency.

2:30 am (Yellow Hammer, Tuesday)

Oral 98 - Engineering organic solvent reverse osmosis performance of vapor phase infiltrated organic-inorganic hybrid membranes

Yi Ren (Georgia Institute of Technology)*

Membrane based organic solvent separations serve as a low-energy alternative to traditional thermal-based separations. While inorganic membranes have high chemical stability to withstand the harsh chemical environments, the high cost and defect rates at scale are obstacles to commercialization. On the other hand, polymer membranes have high scalability with reasonable cost, but the low chemical stability will result in membrane swelling, plasticizing or even complete dissolution in organic solvent environments. Vapor-phase infiltration (VPI) has emerged as a scalable, solvent free post-treatment process that can transform polymer products into organic-inorganic hybrid materials with enhanced properties, including chemical stability desired for membrane based organic solvent separations. The polymeric membrane material is exposed to metal-organic vapor, allowing infiltration of the vapor phase metal-organic molecules into the polymer. The metal-organics will grow into metal-oxide networks that homogeneously mix with polymer chains on a molecular level, prohibiting the polymer chains from large solvent-induced motions, thus hybridizing the entire material into organic-inorganic membrane with higher solvent stability. In this work, the transport of metal-organic vapors into polymer materials will be investigated to better describe and predict the reaction-diffusion transport phenomenon occurring during a typical VPI process for hybridizing polymer membranes. Then, organic solvent reverse osmosis transport of the VPI treated membranes will be studied using an alcohol/alkane mixture system. Finally, organic solvent reverse osmosis transport of the hybridized membrane will be engineered by controlling the metal loading during the VPI process.

3:00 pm (Yellow Hammer, Tuesday)

Oral 99 - Development of a high-performance 30nm NIPS PTFE membrane

Karl Ashkar (Pall Corporation)*; SIAVASH DARVISHMANESH (Pall corporation); Steven Gardner (Pall Corporation)

This abstract highlights Pall Corporation's development of high-performance polytetrafluoroethylene (PTFE) membrane with a mean pore size of 30nm and high organic solvent flux for use in various filtration applications. A proprietary fabrication process involving phase separation was employed to prepare a PTFE phase inversion membrane. The performance of the PTFE membrane was evaluated using a range of test fluids, including organic solvents, and aggressive chemical. The selection of these test fluids was based on their representative nature and relevance to potential filtration applications. The membrane was subjected to a series of characterization techniques, including scanning electron microscopy (SEM) to assess the surface and cross-sectional morphology, and tensile testing, to assess its structural properties and mechanical stability. The membrane was also subjected to various chemicals to determine its solvent stability. Lastly, the membrane was subjected to various thermal conditions to assess its stability at high temperatures. The results showed that the PTFE membrane had excellent filtration efficiency, yielding a mean pore size of 30nm. The membrane also had a high flux rate, indicating low resistance to flow. In addition, the membrane exhibited excellent chemical and thermal stability, and was mechanically robust and stable over a wide pH range and temperatures up to 200C. The 30nm PTFE membrane represents a significant advancement in PTFE phase inversion membrane technology. Its high performance and functional properties make it suitable for a variety of filtration applications, including in the pharmaceutical, microelectronics, organic solvent microfiltration, and chemical industries.

3:30 pm (Yellow Hammer, Tuesday)
Oral 100 - Crosslinked thin film composite membranes for industrial organic complex mixtures separations

Yacine Feliachi (Georgia Tech)*; Ryan Lively (Georgia Tech); M.G. Finn (Georgia Institute of Technology); Ankana Roy (Georgia Tech)

Organic solvent separations, which are commonly used in the chemical, petrochemical, and pharmaceutical industries, are energy-intensive processes. Membrane technology can potentially reduce the energy and resources needed for these separations, but it can be less effective when separating highly similar molecules. Some polymers used in organic solvent reverse osmosis (OSRO) have been known to suffer from plasticization and dilation, making them inefficient for these molecular separations. One solution to this issue is to crosslink the polymer used in the membrane, but traditional methods, such as solvent swelling methods, can be limited by the diffusional movement of the crosslinking agents, and this effect becomes more pronounced as crosslinking agents become bulkier. We have developed a method to introduce larger crosslinking agents to the membrane while overcoming diffusional limitations during the crosslinking infusion step. We utilize the commercial polymer Matrimid® 5218 and four crosslinking agents: para-xylylene diamine, 4,4' biphenyldicarbontrile, spirobifluorene, and triptycene. We introduce a novel Boc protection method to the amines to enable dissolution of the crosslinkers into the membrane casting solution. After casting, the amines within the membrane are thermally deprotected to induce a solid-state crosslinking reaction. The resulting membranes were characterized by FT-IR, gel fraction, helium pycnometry, and TGA. The separations performance was tested in a crossflow OSRO permeation system for a toluene/triisopropyl benzene system and a multicomponent organic solvent mixture. Beyond that, we tested these membranes for performance in a real crude tall oil and crude sulfate turpentine feed. The results showed that the crosslinked TFC membranes had excellent stability in organic solvents and challenging real-world mixtures. Furthermore, the membrane performance could be fine-tuned for separation based on the crosslinking agent and its concentration.

4:00 pm (Yellow Hammer, Tuesday)
Oral 101 - Ultrathin polytriazole membranes for energy-intensive separation

Stefan Chisca (KAUST)*; Suzana Nunes (King Abdullah University of Science and Technology (KAUST))

Climate change is one of the most crucial challenges facing the world today. Simultaneously, fossil-fuel-based energy accounts for more than 80% of the world's primary energy consumption in 2021, suggesting that these industries and the oil refining industry will still play an essential role in the transition towards renewables [1]. However, oil refining is the world's 3rd largest emitter of greenhouse gases [2]. Therefore, there is an urgent need for cleaner and sustainable energy conversion, and membrane technology can be an alternative to conventional energy-intensive separation processes [3]. However, there is still a lack of separation materials operating under industrial conditions and complex mixtures. In addition, highly selective and stable membranes are essential for the success of effective separation processes.

In this work, we show how to combine the properties of new materials, like polytriazole, with conventional membrane fabrication techniques such as phase inversion and thermal crosslinking, to obtain membranes with tunable selectivity for one of the most challenging chemical separations: the fractionation of crude oil [4]. The obtained membranes can discriminate between paraffin over aromatic components. In addition, by tuning the selective layer we can remove sulfur components from the crude oil. The high thermal stability together with tunable selectivity makes the polytriazole membranes a potential candidate to be integrated into hybrid membrane configurations for crude oil fractionation.

Acknowledgment
This research was supported by the King Abdullah University of Science and Technology

References

1. H. Ritchie, et all, OurWorldInData.org, 2021.
2. T. Lei, et all, One Earth, 4, 1114, 2021.
3. S.P. Nunes, P.Z. Culfaz-Emecen, G.Z. Ramon, T. Visser, G.H. Koops, W. Jin, M. Ulbricht, J. Membr. Sci. 598, 2020.
4. S. Chisca, V. Musteata, W. Zhang, S. Vasylevskyi, G. Falca, E. Abou-Hamad, A.H. Emwas, M. Altunkaya, S.P. Nunes, Science, 376, 1105, 2022.

4:30 pm (Yellow Hammer, Tuesday)

Oral 102 - Unveiling the role of sorption, diffusion, interactions and plasticization in organic solvent transport in dense OSN/OSRO membranes

Michele Galizia (University of Oklahoma)*

70% of industrial chemical synthesis occurs in organic solution. Solute concentration and solvent recovery consume approximately 50% of the energy required to produce chemicals and pose problems that are as relevant as the synthesis process itself.

While organic solvent nanofiltration (OSN) and reverse osmosis (OSRO) are emerging as an energy-efficient alternative to distillation, their development is still in its infancy due to the lack of fundamental structure-property correlations which has hampered, for a long time, the design of functional membrane materials suitable for these applications.

In this study, Celazole® polybenzimidazole and PTMSP have been selected as baseline materials to systematically investigate the role of sorption, diffusion, molecular interactions and plasticization on solvent transport. The results show that, contrarily to what has been speculated for many years, sorption governs solvent transport in glassy membranes for organic solvents separation. A detailed molecular mechanism of membrane plasticization was also proposed.

Oral 18 – Membrane Characterization

2:00 pm (Black Warrior, Tuesday)

Oral 103 - Improved Mechanical Compliance in Bicontinuous Cubic Lyotropic Membranes Based on a Cross-Linking Gemini Monomer via Copolymerization with a Non-Cross-Linkable Analog

Lauren N Bodkin (University of Colorado Boulder); Patrick Li (University of Colorado Boulder); Samantha Dyer (University of Colorado Boulder); Zachary Krajnak (University of Colorado Boulder); John Malecha (University of Colorado Boulder); Richard D. Noble (University of Colorado, Boulder)*; Douglas Gin (Department of Chemistry, University of Colorado - Boulder)

Membrane materials for use in chemical protective garments for the U.S. military should have ≥20% reversible stretch without breaking. Copolymerization of a cross-linking gemini lyotropic liquid crystal (LLC) monomer with controlled amounts of a non-cross-linkable analog allows formation of nanoporous, bicontinuous cubic polymer membranes with higher % strain-at-failure values of ≥20% and lower Young's modulus values while retaining high water vapor transport. Although chemical warfare agent (CWA) simulant vapor rejection decreases with the reduced cross-link density, increased mechanical pliability was achieved, demonstrating that the mechanical properties of these LLC-based, breathable CWA-barrier materials can be modified in this fashion.

2:30 pm (Black Warrior, Tuesday)

Oral 104 - Advanced metrology, molecular dynamics, and machine learning for determining ion activity coefficients in ion-exchange membranes

Christopher G Arges (The Pennsylvania State University)*; Hishara Keshani Gallange Dona (Louisiana State University); Teslim Olayiwola (Louisiana State University); Luis Briceno-Mena (Louisiana State University); Mario Ramos-Garces (The Pennsylvania State University); Revati Kumar (Louisiana State University); Jose Romagnoli (Louisiana State University)

The activity coefficients of ions in polymeric ion-exchange membranes (IEMs) dictate the equilibrium partitioning coefficient of the ions between the membrane and the liquid and it also affects ion transport processes such as migration and diffusion. The permselectivity of an ion-exchange membrane for counterion type i over counterion type j is related to the product of the partitioning coefficient and ratio of ionic mobility coefficients (i.e., a form of the solubility-diffusivity model). Accurately predicting the ion activity coefficient without experimental data has been elusive as most activity coefficient models are empirical or semi-empirical. This talk commences with a multitude of experimental methods for determining the ion activity coefficient in IEM materials. The second part of this talk discusses our recent work using molecular dynamics (MD) and machine learning (ML). The MD-ML framework uses small experimental activity coefficient datasets in conjunction with polymer structure information and molecular attributes describing the solvation of ions and polymers to predict the ion activity coefficient in IEMs. Two different ML models were developed to estimate the molecular attributes and the ion activity coefficient. The best ML model accurately predicts the solvation descriptors and ion activity coefficient with an average mean absolute error of <7% and 10%, respectively. Adopting the said approach allow for the estimation of ion activity coefficients in IEMs without the need for new time-consuming MD simulation runs and lengthly experiments.

3:00 pm (Black Warrior, Tuesday)
Oral 105 - Mathematical Modeling of CO2 Facilitated Transport across Amine-functionalized Polymeric Membranes with Direct Operando Observation of Amine Carrier Saturation

Casey O'Brien (University of Notre Dame)*

Reactive amines in polymeric facilitated transport membranes facilitate CO2 transport across the membrane to give rapid and selective CO2 separation from mixed gas streams. However, the fundamental kinetics and mechanisms of amine-facilitated CO2 transport are currently not well understood on the molecular level. In this talk, I will discuss a new mathematical model that we have used to extract fundamental kinetic and thermodynamic parameters associated with CO2 facilitated transport, including the permeability of CO2 facilitated transport, the permeability of CO2 solution-diffusion, and the equilibrium constant associated with the interaction of CO2 with the amine group. The model was fit to experimental CO2 fluxes measured across polyvinylamine (PVAm_ as a function of CO2 partial pressure (2-100 kPa) and PVAm thickness (1-4 μm). Our results indicate that the permeability of CO2 via facilitated transport of carbamate is an order of magnitude greater than the permeability of CO2 via solution-diffusion. The model also shows the amine carriers become saturated at a CO2 partial pressure of ~10-15 kPa, and the CO2 flux across the membrane occurs primarily via solution diffusion at pressures greater than this. This amine carrier saturation was directly observed—for the first time to our knowledge—at precisely this CO2 partial pressure range using operando Raman spectroscopy and FTIR spectroscopy techniques developed in our lab that can detect the formation of carbamate species and simultaneously measure CO2 fluxes across the membrane. Thus, the mathematical model is in good agreement with spectroscopic observations of CO2 interactions with PVAm. The molecular-level insights obtained from this new kinetic modeling and operando methodology will enable a more rational design of membranes for CO2 separation than conventional membrane testing with ex-situ characterization approaches.

3:30 pm (Black Warrior, Tuesday)
Oral 106 - Unveiling the mechanism of light gas transport through configurational free volume in glassy polymers

William Box (University of Oklahoma)*; Zihan Huang (University of Notre Dame); Ruilan Guo (University of Notre Dame); Michele Galizia (University of Oklahoma)

Conventional glassy polymers suffer from three major issues, the permeability-selectivity trade-off, physical aging, and plasticization. These issues represent a roadblock to progress in the field and hamper the design of membranes exhibiting enhanced selectivity and stability. Polymers containing iptycene units, that is, configurational free volume (CFV), help overcome this limitation. The internal free volume of triptycenes is not tied to molecular conformation, as the excess free volume in conventional glassy polymers is. Moreover, the internal volume size of iptycene units is extremely regular and comparable to the size of a single molecule, making iptycene-based polymers highly selective. However, the fundamental mechanism of small molecule transport in polymers exhibiting CFV is poorly understood. To fill this knowledge gap, a class of polybenzoxazoles (TPBOs) with systematically varied amounts of CFV were selected to perform a light gas (i.e., N2, CH4, CO2) transport study, and propose a possible mechanism of small molecule transport in polymers exhibiting CFV.

Solubility and solubility-selectivity were found to be virtually unaffected by CFV. In contrast, CFV affects diffusivity and diffusivity-selectivity (αD) to a significant extent. Henry and Langmuir mode contributions to overall diffusion were then separated to understand their roles in CO2/CH4 αD. Remarkably, while the Henry αD is fairly constant with increasing CFV, Langmuir αD increases to infinity at a triptycene concentration of 75% or higher, demonstrating that larger molecules (CH4) become increasingly hindered compared to smaller molecules (CO2) when more CFV is incorporated in the polymer. Therefore, our study provides the final proof of the superior size-sieving ability exhibited by iptycene-based polymers by unveiling, for the first time, its molecular origin. Finally, an isosteric heats analysis of CO2 sorption provides additional information regarding TPBOs plasticization propensity.

4:00 pm (Black Warrior, Tuesday)
Oral 107 - Characterizing dielectric properties and water molecule dynamics using dielectric relaxation spectroscopy
Sean M Bannon (University of Virginia)*; Charles Leroux (University of Virginia); Beatrice M Tremblay (University of Virginia); Geoffrey M Geise (University of Virginia)

Ion transport in polymeric membranes is governed by interactions between the polymer and sorbed water and ions. Broadband microwave dielectric relaxation spectroscopy (DRS) is a tool that contributes to characterizing these interactions in hydrated polymers by measuring dielectric permittivity properties, e.g., the static permittivity (dielectric constant) and water molecule dynamics (related to the dielectric loss), which can help to inform ion transport properties. Here, we discuss broadband microwave (i.e., 45 MHz – 25 GHz) DRS measurements made on hydrated sulfonated poly(aryl ether sulfones) and research focused on understanding how polymer structure (e.g., chemical functionality and degree of sulfonation) affects relative permittivity properties.
Characterization using DRS may also aid efforts to model ion transport in polymers. Many models used to describe ionic interactions in polymers, such as Manning's model for counter-ion condensation, require information about the hydration-dependent dielectric constant of the polymer. In the absence of polymer specific data, a linear volume additivity-based approach, which may work well for Nafion, has been used to estimate the dielectric constant of hydrated polymers based on water content alone. However, recent DRS measurements suggest that the dielectric constant of many hydrated hydrocarbon polymers does not depend on water content in the same manner as Nafion, and for some polymers, the volume additivity-based approach over-predicts the hydrated polymer's dielectric constant by an order of magnitude. This presentation will also discuss analysis of the dielectric constant using the Maxwell-Garnett model to describe the relationship between polymer water content and dielectric constant in several hydrated polymers.

4:30 pm (Black Warrior, Tuesday)
Oral 108 - A Comprehensive OMICs Platform for Detecting Biofouling in Full-Scale Reverse Osmosis Facilitates
Weiliang Bai (University of Texas at Austin)*; Manish Kumar (University of Texas at Austin)

Reverse Osmosis (RO) is widely used in water reuse and desalination plants due to its high permselectivity and mild temperature requirement. The formation of biofouling limits the lifetime of RO membranes because the colonized biofilm decreases the permeability and rejection rate. Understanding the composition of biofilms attaching to prefilters and RO membranes helps designing targeted preventive measurement and the selection of the optimal period for cleaning or replacement.
In this study, an OMIC platform was applied to analyze the taxa diversity in feed, prefilter, and RO membranes in the two of the largest RO plants in the U.S, the Claude "Bud" Lewis Carlsbad Desalination Plant and the Tampa Bay Seater Desalination facility. Distinct microorganisms were found in the prefilter and the RO membranes demonstrating the cartridge prefilter could retain some cells but not all. Some opportunistic bacteria which could pass through the prefilter and grow on the membrane surface may be the principal factor for biofouling. The complexity of microorganisms over the two RO desalination plants proposed the necessity of building a more comprehensive monitoring system to understand the transformation of colonized microbes on RO membranes.
To achieve time-series monitoring of the biomass attached to the membrane surface, we built several membrane fouling simulators (MFS) in parallel with the RO membrane stages at Orange County Water District. The DNA analysis has been employed for RO membrane stage, and MFSs. The relative taxonomic abundance is similar for the RO membrane and the MFSs. The result from Beta diversity PCoA revealed the microorganisms from MFS were similar to the corresponding RO membrane. This similarity supported that easy-sampled MFS is capable of monitoring and representing the biofilm on the membrane surface. The analysis result from time-series MFSs samples provided an insight of how the structure of biofilm transformed on the membrane surface.

Oral 19 – Electrochemical Separations 2

2:00 pm (Capstone, Tuesday)

Oral 109 - A new polymer platform for anion exchange membranes based on Friedel-Crafts polymerization

Lingyu Yang (University of Notre Dame); Chengkai Fan (University of Notre Dame); Haifeng Gao (University of Notre Dame); Jennifer L Schaefer (University of Notre Dame)*

Anion exchange membrane fuel cells are recognized as promising energy conversion devices due to their attractive advantages in basic operating conditions. However, the greater application is still impeded by the poor stability and low hydroxide conductivity of anion exchange membranes in alkaline conditions. Here we report on a tunable AEM platform based on Friedel-Crafts (F-C) polymerization for modulation of ionic conductivity, mechanical properties, and alkaline stability. Structure-property relationships between polymer sequence, mesoscale structure, ionic conductivity, mechanical properties, and alkaline stability are elucidated.

2:30 pm (Capstone, Tuesday)

Oral 110 - Highly Charged Ion-Exchange Membranes for Treating Concentrated Salt Solutions via Electrodialysis

Jovan Kamcev (University of Michigan)*

Brine management is an enormous challenge for the treatment of nontraditional water sources such as brackish water, seawater, municipal and industrial wastewater (e.g., produced water), especially in inland areas where brine disposal options are limited. Electrodialysis (ED) is being considered as a lower energy alternative to thermal brine concentration and could enable cost-effective treatment of brines via concentration and subsequent crystallization. However, conventional ion-exchange membranes (IEMs), originally designed for brackish water desalination, exhibit low permselectivity when contacted by highly concentrated salt solutions, making them unsuitable for this application. While new classes of commercial IEMs for brine treatment via ED have emerged, further improvements in membrane performance are necessary for ED to outcompete conventional technologies in this space. Rational design of better performing IEMs would be facilitated by improved fundamental understanding of the connection between polymer structure and transport properties. In this presentation, I will discuss the development of highly charged IEMs for brine concentration via ED, with a specific focus on understanding their transport properties when contacted by concentrated salt solutions. I will also discuss some practical considerations related to designing new membranes for this application, including fabricating mechanically reinforced IEMs and bench-scale electrodialysis tests.

3:00 pm (Capstone, Tuesday)

Oral 111 - Production of polyphenols-enriched cranberry juice through electrodialysis with filtration membrane: How physicochemical properties of filtration membranes impact anthocyanins migration and selectivity during electrodialysis

Eva Revellat (Université Laval; Institute of Nutrition and Functional Foods)*; Laurent Bazinet (Laval University, Institute on Nutrition and Functional Food (INAF))

Health-promoting effects credited to anthocyanins have been extensively studied and recent discoveries have suggested their impacts on the modulation of the gut microbiota . Cranberry juice - a goldmine of anthocyanins - could be enriched to enhance its heath benefit. In previous studies, electrodialysis with filtration membrane (EDFM) selectively enriched juice in anthocyanins. The present study focused on this green and promising technology especially on understanding how physicochemical properties of filtration membranes impact global and selective anthocyanin migration during EDFM. Thus, six polyether sulfone (PES) and polyvinylidene fluoride (PVDF) membranes with molecular weight cut-offs (MWCO) of 150kDa, 200kDa, 250kDa, 300kDa and 500kDa were characterized and used during EDFM treatment. Membranes were characterized in terms of thickness, conductivity, contact angle, % of hydrophilic pores, porosity, zeta-potential, roughness. Juices were analysed using UV-visible spectroscopy to assess the global anthocyanins enrichment and HPLC was used to determine the individual anthocyanins, proanthocyanidins and organic acids migration. Results have shown an impact of MWCO and membrane material on anthocyanin's enrichment. Data are currently analysed by redundancy analysis (RDA) and multivariate regression to understand the correlation between membrane physicochemical properties and global and individual anthocyanins migration during EDFM : migration predictive models will be generated.

3:30 pm (Capstone, Tuesday)

Oral 112 - Correlation between volatile fatty acid transport properties in anion exchange membranes and volatile fatty acid transport during electrodialysis

Holly M Haflich (University of North Carolina- Chapel Hill)*; Josh Singleton (University of North Carolina- Chapel Hill); Orlando Coronell (University of North Carolina at Chapel Hill)

Electrodialysis (ED) electrically transports charged species through oppositely charged ion exchange membranes. ED can recover economically valuable volatile fatty acids (VFAs) from biological and waste mixtures given the anionic charge of VFAs. Solute transport properties in the membrane (i.e., partition coefficient (K), diffusion coefficient (D), and permeability (P=KD)) affect solute flux in ED and are dependent on solute and membrane physico-chemical properties. Therefore, to advance fundamental understanding of VFA flux through anion exchange membranes (AEMs) during ED, our objective was to elucidate the relative contributions of partitioning and diffusivity (mobility) of VFAs and inorganic anions to their flux during ED. Accordingly, for a group of VFAs (C2-C5) and inorganic anions (Cl-, NO3-), we experimentally determined the anion fluxes in ED, partition coefficients, and permeabilities, and calculated the diffusion coefficients from P and K. We then evaluated the correlations between anion flux and transport properties (P, K, D). Results showed permeability had the strongest correlation with anion flux (R2=0.98, p<0.01), consistent with flux described by the Nernst-Planck equation. Further, flux had a strong correlation with D but only a weak correlation with K, indicating that differences between anion fluxes were more strongly influenced by differences in mobility than partitioning. We observed a negative correlation between the VFA diffusion coefficient and VFA carbon chain length which is consistent with expectations from increased drag and steric hindrance with greater size. The transport of longer-chain VFAs was further hindered by increased deswelling of the AEM with VFA chain length. This work provides fundamental insight on interactions between anions (VFAs, Cl-, and NO3-) and AEMs which affect anion flux during ED. Future work will investigate the impact of solute transport properties on selective VFA and inorganic anion transport during ED.

4:00 pm (Capstone, Tuesday)

Oral 54 - Bipolar membrane electrodialysis for direct LiOH production from concentrated brines

Aubrey Quigley (The University of Texas at Austin)*; Kevin Reimund (University of Texas at Austin); Nathaniel Lynd (The University of Texas at Austin); Benny Freeman (The University of Texas at Austin)

Recent developments in the electric vehicle market necessitate more efficient means of lithium extraction as market demand is projected to rapidly outpace current production. Currently, lithium production from brines involves lengthy evaporation and processing steps and yields Li2CO3 rather than LiOH, the preferred form for production of advanced battery cathode materials. An emerging field of study in direct lithium extraction from brine mixtures is bipolar membrane electrodialysis (BPED) which employs an electrodialysis membrane stack to directly produce LiOH from a Li-containing saline feed without the employment of pretreatment steps. However, most commercial cation exchange membranes do not possess specific solute selectivity. Calcium and magnesium are of particular concern due to their prevalence in common brines and their tendency to produce insoluble hydroxides (e.g., Ca(OH)2 and Mg(OH)2) in the base product. These hydroxides can foul and damage the membrane surface, leading to an increase in the overall stack resistance. Thus, current cation exchange membranes are limited to applications where the feed concentration of divalent cations is below 10 ppm. In this work, we synthesized a selective cation exchange membrane for use in BPED which inhibits the transport of divalent cation species. In mixed-salt BPED experiments, the selectivity of lithium over calcium and magnesium was measured over a range of divalent species concentrations showing relative transport numbers of lithium greater than 100. The selective cation exchange membranes we have developed can be used on feeds which have not yet been depleted of divalents, decreasing the number of pretreatment steps for direct electrochemical production of LiOH from highly concentrated brines.

4:30 pm (Capstone, Tuesday)

Oral 114 - Revolutionizing Ion Transport: Dynamic Channel-Size Tuning in MXene Transistor Membranes for Selective Separation and Enhanced Control

Aaditya A Pendse (Lawrence Livermore National Labortory)*; Aleksandr Noy (Lawrence Livermore National Laboratory)

Ion transport is critical for many applications such as mass delivery, energy conversion and bioelectronics. 2D materials are excellent candidates for synthetic membranes that display outstanding ion separation and transport properties. However, maintaining high ion flow rates while selectively excluding specific ions remains a challenge. In this study, we present a novel approach to tuning the interlayer spacing of MXene using an external gate voltage to control ion transport.

MXene materials have high electronic conductivity, surface functionality and hydrophilic channels, making them stable in water. These negatively charged membranes have been extensively used for ion separations, and previous research has demonstrated a statically-tunable interlayer spacing via ion intercalation, resulting in selective transport of metal ions of varying size and charge. However, the selectivity is limited to a characteristic ionic size.

Our study couples MXene membranes with electronic gating approaches to dynamically adjust the membrane medium for efficient ion separations. We evaluate transport behavior, ion selectivity, and gating performance, demonstrating that cations could move thermodynamically uphill in response to the externally applied voltage bias over a broad range of concentrations, at rates much faster than those via simple diffusion.

We also demonstrate precise control over the membrane interlayer spacing in response to varying gate voltage as a result of charge potential regulation at the MXene surface. In addition, we establish a quantitative relation between channel-size and gate voltage using in-situ XRD studies, revealing a membrane system capable of dynamic membrane tuning to selectively enhance or inhibit the transport of specific ions. Overall, this study presents a promising approach to address the challenges associated with selective ion transport and paves the way for the development of novel separation platforms with many potential applications.

Oral 20 – Membrane Distillation and Pervaporation

2:00 pm (Private Dining, Tuesday)

Oral 115 - Anti-scalants for mitigating silica scaling in membrane desalination: Effects of molecular structure and membrane process

Yiqun Yao (Colorado State University)*; Tiezheng Tong (Colorado State University)

Membrane-based desalination and wastewater treatment are promising strategies that enable water purification from various unconventional resources. However, mineral scaling constrains the performance and efficiency of membrane technologies significantly. Silica (SiO_2) is a major mineral scale that is formed via a polymerization reaction. Due to its amorphous nature and the lack of effective anti-scalants, silica scaling is considered as the Gordian Knot of membrane desalination processes. In this work, we performed a systematic study that investigated the design principle and mechanism of anti-scalants for mitigating silica scaling in membrane desalination. We examined the effectiveness of anti-scalant candidates with different molecular functionalities in hindering silica scale formation in static experiment as well as in dynamic crossflow membrane distillation (MD) and reverse osmosis (RO) experiments. Particularly, I will present three key results that answer the following questions: (1) how the anti-scalant performance behaves as a function of molecular functionality and size; (2) whether the efficiencies of anti-scalants revealed in static experiments predict those in dynamic, crossflow membrane filtration; and (3) whether the anti-scalants possess the same efficiencies in RO and MD. Our findings provide fundamental insights that have the potential to guide successful design of anti-scalants tailored to silica scaling for membrane desalination.

2:30 pm (Private Dining Tuesday)

Oral 116 - New approaches to mitigate wetting and scaling in membrane distillation

Shihong Lin (Vanderbilt University)*

Membrane distillation (MD) with hydrophobic membrane has been known to be susceptible to fouling, wetting, and scaling. For the past decade, efforts have been dedicated to developing novel membranes with special wetting properties for mitigating these challenges. In this talk, I will first provide a summary of how novel membranes (omniphobic, superhydrophobic, and Janus) were used to control wetting, scaling, and fouling, as well as the limitations of those approaches. I will then discuss two recent developments in improving the robustness of MD. The first is Janus membrane with dense hydrophilic coating, which has been well known for its fouling resistance but was only recently identified as highly effective for wetting mitigation. The second is the operation of MD under negative gauge pressure, which is a simple yet highly effective approach that can alleviate both scaling and wetting with commercial hydrophobic membranes. These two approaches feature simplicity, multifunctionality, and practicality, and demonstrate strong potentials to advance MD toward practical applications.

3:00 pm (Private Dining, Tuesday)

Oral 117 - Hydrophobic Hollow Fiber-Membrane for Spacesuit Cooling

Dibakar Bhattacharyya (U. Kentucky)*; M.Arif Khan (University of Kentucky); Glenn Lipscomb (University of Toledo, Toledo); Andrew Lin (University of Kentucky); Morgan Abney (NASA Langley Research Center)

The large amount of literature published on membrane distillation (particularly VMD) applications using hollow fiber membranes, is relevant to the design of innovative devices like NASA's hollow fiber-based Spacesuit Water Membrane Evaporator (SWME), The SWME will be a critical component of the portable life support system in spacesuits used for extravehicular activities. It is designed to reject the heat generated by the crew member wearing the suit and the electrical components of the suit's portable life support subsystem. The rate of water evaporation and associated exiting liquid water temperature are controlled by adjusting the water vapor pressure in the shell of the hollow fiber module with a backpressure valve. In contrast to VMD, SWME operations is at a lower inlet temperature and the main goal is to cool the water in the lumen side rather than producing a particular quality distillate, and no vacuum pump is needed since MARS is at 7 torr and Moon at zero torr. The HF PP membrane used here has lower pore diameter (about 40 nm), and thus high LEP.. This work develops a comprehensive predictive mathematical model to estimate the thermal performance of the NASA spacesuit water membrane evaporator (SWME). The model is based on mass and energy balances in terms of the physical properties of water and membrane transport properties. The mass flux of water vapor through the pores is calculated based on Knudsen diffusion with a membrane structure parameter (SP) that accounts for effective mean pore diameter, porosity, thickness, and tortuosity. Lumen-side convective heat transfer coefficients are calculated from laminar flow boundary layer theory using the Nusselt correlation. NASA reported experimental thermal performance data was used to fit the model and estimate SP to be 1.04 x10-4. After validation, the model was used to perform a detailed sensitivity analysis to study the effect of input variables changes on the output thermal performance parameters.

3:30 pm (Private Dining, Tuesday)

Oral 118 - Radiofrequency responsive membranes for membrane distillation applications

Md Hasib Al Mahbub (Texas Tech University); Mohammad Saed (Texas Tech University); Mahdi Malmali (Texas Tech University)*

Membrane Distillation (MD) is regarded as a promising and relatively new separation process technology for the desalination of hypersaline water (>70 g/L). MD process involves a microporous hydrophobic membrane which acts as a barrier to the liquid brine phase passed at an elevated temperature, while only the vapor phase transport through the membrane pores. The driving force for the MD separation process is the temperature gradient across the membrane which induces a partial pressure gradient. The vapor phase which is mostly water vapor is then condensed and collected as pure water on the permeate side. This work reports the design of a novel energy-efficient radio frequency (RF) heated MD process, in which a conductive hydrophobic membrane was used to heat the membrane surface for intensified mass transport. The conductive hydrophobic membrane was fabricated by laser writing on a polyethersulfone (PES) membrane, which converted the polymeric substrate into graphene. PES-LIG membranes were characterized for Raman spectroscopy, SEM-EDX, and XPS analysis, where the laser-ablated graphene was confirmed on the PES surface without destroying the porous structure of the membrane. PES-LIG surface hydrophobicity was studied from water contact angles, which increase from 60 (for pristine PES) to 143 degrees.
The LIG-PES membranes fabricated using this method were demonstrated for lab-scale vacuum membrane distillation (VMD), by only surface heating of the graphene layer on the membrane surface using radiofrequency (RF) electromagnetic waves. An RF applicator was designed and placed inside the membrane cell to heat the membrane surface. The thermal response of PES-LIG membranes was recorded with a FLIR camera under an electromagnetic field to find the optimum RF parameters (power and frequency) to investigate the VMD performance. Membrane surface temperature under the RF field was found to be in the range of 90 to 200 degC for RF frequency in the range of 90 to 200 MHz.

4:30 pm (Private Dining, Tuesday)

Oral 119 - Materials and design to enhance transport in membrane distillation

Waqas Alam (Purdue)*, David M Warsinger (Purdue)

Thermal desalination processes like membrane distillation (MD) can reap substantial benefits from advanced heat transfer techniques, especially because energy is most of MD lifetime costs. This process, which resembles a heat exchanger with an additional membrane layer, has transport needs that differ from other heat exchangers because membrane distillation desires to maximize mass transfer to provide fresh water. This focus of mass transfer instead of heat transfer means simultaneously minimizing some modes of heat transfer, especially conduction through the hydrophobic membranes and adjacent air. Here, we cover the benefits and tradeoffs of a wide variety of thermal enhancements, including jumping droplet, dropwise, and SLIPS condensation, porous condensers, and conductive spacers. Furthermore, we show how multilayer membranes can be designed for maximizes mass transfer while minimizing undesirable modes of heat transfer. Numerical modeling and experimental results show that large efficiency enhancements, even doubling efficiency, is possible over state-of-the art systems using enhanced materials, including copper wicks, and superhydrophobic nanostructures. Similar approaches can be applied to many other thermal desalination technologies to enhance their viability and provide fresh water and recover resources from brines.

4:30 pm (Private Dining, Tuesday)

Oral 120 - Omniphobic electrospun membranes for treatment of hydraulic fracturing produced water through membrane distillation

Ranil Wickramasinghe (University of Arkansas)*; YuHsuan Chiao (University of Arkansas)

Maximizing reuse of wastewater is essential to meet increasing demands for fresh water. Hydraulic fracturing oil and gas flow back and produced water is frequently highly impaired. Currently it is often deep well injected into geologically isolated formations. Recovery of this water for beneficial uses is essential to promote a circular economy. While membrane distillation could be used to recover water from high salinity wastewaters, membrane fouling limits its practical application. Here a superomniphobic polyvinylidene fluoride (PVDF-F) membrane was fabricated through electronspinning to obtain an excellent antiwetting and antifouling property for the effective treatment of produced water. Compared with pristine PVDF and hydroxyl functionalized PVDF (PVDF-O) membrane, PVDF-F membrane had a larger pore diameter, while its pure water flux was found to be lower than both. Neither liquids with high surface tension (water, ethanol) or low surface tension (sodium dodecyl sulfate (SDS) as surfactant, mineral oil) could wet the PVDF-F surface even after 3 min of contact. During membrane distillation (MD) of 3.5 wt% of aqueous NaCl solution, PVDF-F could withstand 0.4 mM SDS solution without any wetting as seen from the conductivity of the permeate, which was not the case for the PVDF and PVDF-O membranes. After 18 h of MD operation using produced water, a significantly thick fouling layer was found on the PVDF and PVDF-O membranes containing Ca3(PO4)2, Sr3(PO4)2 and Mg3(PO4)2. However, PVDF-F had a very thin and reversible organic fouled layer, causing only 40% reduction in permeate flux and with a low permeate conductivity of 20 μS cm−1. This layer was easily removed by cleaning using distilled water and 80% of the flux was recovered. On the 2nd and 3rd MD cycle, the reduction in permeate flux was within 10%. The PVDF-F membrane was run for three successful consecutive MD cycles during produced water processing without any significant fouling as well as wetting.

Abstracts – Oral Presentations, Wednesday

Presenting authors are indicated by an asterisk*.

Oral 21 – NAMS Awards Session

9:30 am (Druid, Wednesday)

Oral 121 - Challenges and opportunities in ultra-high pressure reverse osmosis operation

Jishan Wu (UCLA)*; Eric Hoek (UCLA)

Herein, we evaluate the performance (i.e., flux and rejection) of commercially-available, thin film composite brackish water RO (BWRO), seawater RO (SWRO) and high-pressure RO (HPRO) membranes operating at pressures from 14 to 207 bar (200 to 3,000 psi). For each membrane material, we elucidate the impacts of permeate carrier morphology on performance using a porous metal frit and woven tricot mesh materials extracted from commercial HPRO, SWRO, BWRO and tap water RO (TWRO) membrane modules. The water permeability of all tested membranes declines with increasing pressure, whereas rejection behaves differently for different combinations of membrane type and permeate carrier. Cross-sectional SEM and FIB-SEM images confirm permanent reduction of the polysulfone support membrane thickness (38% to 60%) as well as collapse of support membrane skin layer pores – both of which contribute to the observed performance decline. Also, at ultra-high pressures, permeate carrier materials with higher porosity caused greater embossing, and ultimately, coating film damage (defect formation) that leads to increased salt passage (loss of salt rejection). In contrast, the permeate carriers with lower porosity still lost water permeability, but maintained higher rejection. Finally, all of the observed compaction and embossing-related performance decline occur within about 60 minutes after a membrane coupon was exposed to ultra-high pressure. Furthermore, we will present a new RO membrane construction with superb compaction-resistance operating up to 207 bar feed pressure. Cross-sectional SEM confirms less than 10% thickness reduction (compaction) of the membrane. In addition to superior compaction-resistance, the novel RO membrane produces good water permeability and rejections over 99% to NaCl solutions (concentrations up to 200,000 mg/L).

10:00 am (Druid, Wednesday)

Oral 122 - Kinetic Control of Angstrom-scale Porosity in Graphene for Direct Scalable Synthesis of Atomically Thin Proton Exchange Membranes

Nicole K Moehring (Vanderbilt University)*

Proton selective defects in the lattice of atomically thin 2D materials such as graphene which allow proton flux higher than pristine materials while maintaining selectivity could enable their use as membranes for energy conversion/storage applications if scalable synthesis techniques are utilized. Here, we show for the first time that kinetic control of graphene synthesis during scalable, chemical vapor deposition can allow for formation of angstrom-scale proton selective pores and facile fabrication of large-area atomically thin proton exchange membranes. Systematic liquid and gas phase transport measurements on the same membrane paired with a resistance-based transport model, we study the influence of these intrinsic pores on selective proton transport through centimeter-scale Nafion|Graphene|Nafion membranes. Further, we establish a novel membrane fabrication approach that effectively mitigates transport of undesired species (small ions and H2 gas crossover) while maintaining proton transport suitable for applications. Our insights on kinetic control of angstrom-scale pore formation over large areas during synthesis as well as the facile approach for membrane fabrication offer new avenues to enable functional, large-area, atomically thin proton exchange membranes.

10:30 am (Druid, Wednesday)

Oral 123 - Data-driven predictions of complex mixture permeation in polymer membranes

Youngjoo Lee (Georgia Tech)*; Lihua Chen (Georgia Tech); Rampi Ramprasad (); Ryan Lively (Georgia Tech)

Separations of nonaqueous or organic-water liquid mixtures by polymer membranes have been identified as a key enabler for reduced carbon emissions in chemical, biochemical, and petrochemical manufacturing processes. The fractionation by membranes is attractive when hybridized with existing separation modalities (e.g., distillations) for hydrocarbon separations, and has the potential to be applied to biobased complex mixtures (e.g., biocrude oils, crude tall oils) that are vulnerable to high temperature operations. A critical gap for this class of membrane separations is the difficulty associated with estimating membrane performance when challenged with a complex mixture. Current approaches rely on laborious and specialized experiment that requires sophisticated and time-intensive analysis to understand the effectiveness of the membrane. To enable rapid and quantitative predictions of the separation of any arbitrary complex mixture using any arbitrary linear polymer membrane, we developed machine learning algorithms capable of predicting guest diffusivities and solubilities based on the chemical structures of the polymer and the solvents in a mixture. These transport properties are parameterized in a Maxwell-Stefan approach to calculate the flux of each molecule in a complex mixture. The entire prediction framework was validated by several separation measurements from binary liquid mixtures to significantly complex mixtures such as real crude oils consisting of 60,000 molecules. The model accurately predicted the fluxes and compositions of permeates from the crude oil fractionation measurements within 6-7 %. This work will open new opportunities to efficiently explore new membrane materials and separation processes not with any experimental pre-inspection but only with easily accessible information of the polymer and mixture. Details of the future direction and potential applications of this work will be discussed.

11:00 am (Druid, Wednesday)

Oral 124 - Modification and Synthesis of Polyamide Reverse Osmosis Membranes for the Separation of Small, Neutral Molecules

Shahriar Habib (The University of Alabama); Madison A Wilkins (The University of Alabama); Bryn Larson (The University of Alabama); Steven T Weinman (The University of Alabama)*

Reverse osmosis (RO) membranes have been the gold standard for water desalination for 30 years. Even though RO membranes exhibit excellent performance rejecting monovalent and divalent salt ions, they do not reject small, neutral molecules (SNs) at adequate levels, especially at near-neutral pH. Due to the fast, uncontrolled nature of the interfacial polymerization reaction, the polyamide layer contains both network and aggregate free volume holes (pores). Because SNs are not affected by the charge exclusion rejection mechanism that allows for high salt rejection, reducing the free volume to reduce the passage of the SNs through the membrane is needed. The available free volume is a function of the monomers and polymerization conditions used to synthesize these polyamide layers. In this regard, the modification of RO membranes to increase the degree of cross-linking and altering the interfacial polymerization synthesis process of polyamide layers with new surfactants are ideal approaches. We hypothesize that if the polyamide layer can be further cross-linked with small linear diamines possessing free rotating ability, then the polyamide layer free volume hole size will reduce to increase SN rejection. Additionally, we hypothesize that surfactants provide control of the IP reaction to achieve a homogenous polyamide layer with sharp selectivity without sacrificing water permeance. In this work, we modified polyamide layers of commercial RO membranes using different concentrations of m-phenylenediamine (MPD) and numerous linear diamines. Also, we synthesized traditional RO polyamide layers using new surfactants. We characterized the diffusion rate of MPD into the organic phase when changing the surfactant and concentration of surfactant. The membranes were characterized using SEM, XPS, AFM, contact angle goniometry, zeta potential, and ATR-FTIR. Membranes were performance tested for water permeance and NaCl, urea, and boric acid rejection using a dead-end stirred cell.

11:30 am (Druid, Wednesday)

Oral 125 - High-Performance Non-Aging Rubbery Thin Film Composite Membranes for Post-Combustion CO2 Capture: From Material Synthesis to Field Demonstration

Lingxiang Zhu (National Energy Technology Laboratory)*; Victor A Kusuma (NETL); David Hopkinson (NETL)

The U.S. government is targeting a net-zero carbon-emission economy by 2050, offering an exciting opportunity for membrane-based post-combustion CO_2 capture from industrial point sources. Given that industrial flue gas (mainly a CO_2-N_2 mixture) has low CO_2 partial pressures (0.05-0.3 bar) and high volumetric flowrates, high-permeance and CO_2-selective membranes are needed to make membrane technology an economically viable option for large-scale deployment. Thin film composite (TFC) membranes, comprised of a thin selective layer and a porous support layer, are necessary for practical implementation because they can provide high permeance while minimizing the use of selective materials. Currently, most studies of TFC membranes for CO_2/N_2 separation focus on developing ultra-high permeability selective materials while relying on commercial ultrafiltration membranes as a membrane support. These material combinations encounter several challenges in realization as high-permeance TFCs, such as accelerated physical aging of PIM-based thin selective layers, difficulties in forming defect-free mixed-matrix thin films, and significant mass transfer resistance imposed by membrane supports. To overcome all of those issues, this study reports the rational design and fabrication of a highly permeable non-aging TFC membrane achieved by: (1) synthesizing a high-performance rubbery polyphosphazene-based selective material; (2) developing a high-porosity membrane support; and (3) optimizing coating methods to assemble the two materials into scalable membranes. The resulting membrane yields remarkably high CO_2 permeance of 4,500 GPU and CO_2/N_2 selectivity of 34, and more importantly, there is no reduction in gas permeance when challenged with a 1000-hour physical aging test. This presentation will cover a full spectrum of this membrane R&D including material synthesis, membrane fabrication and scale-up, and field demonstrations at the National Carbon Capture Center and a steel mill.

12:00 pm (Druid, Wednesday)

Oral 126 - Tuning 2D materials channels for sub-nano scale separation: from interlayers to in-plane pores

Jie Shen (King Abdullah University of Science and Technology)*

2D materials with atomic thickness and large lateral dimensions offer great opportunities to create membranes that can theoretically afford ultimate permeation properties. However, the practical development of 2D materials membranes for sub-nm scale separations is hindered by some critical challenges, including the controlling of non-selective defects and channel size distribution, and long-term stability under industrial-like conditions.

In this talk, I will present our strategies for designing and tuning 2D materials channels and their mass transport behaviors at the sub-nm scale, including selective transport of gas, liquids, and ions as separation membranes. We developed a polymer-induced assembly strategy to reduce the non-selective defects in the membranes and enabled fast and selective CO2 or H2 transport through the 2D interlayer channels (Angew. Chem. Int. Ed., 2015, 54, 578; ACS Nano 2016, 10, 3398). To control the channel size and microenvironment, we developed approaches including cationic control of interlayer spacing (Nature 2017, 550, 380) and covalent functionalization of 2D channels (Adv. Funct. Mater. 2018, 28, 1801511), which improved the membrane performances of water/ion separation and CO2 separation, respectively. We also intercalated 1D polymer nanowires acting as "nano-bridges" between the 2D layers to significantly increase the interlayer binding and mechanical strength of the membranes towards practical applications (Sci. Adv. 2020, 6, eabb5367). In contrast to the previous efforts that uncontrollably drilled holes in 2D thin layers, we fabricated nanometer-thick conjugated-polymer-framework membranes (Nat. Mater. 2022, 21, 1183) and monolayer transition metal dichalcogenide membranes (to be submitted), which intrinsically had sub-nm in-plane pores that were robust and highly selective to water. The long-term stability and antifouling properties of the membranes were also studied.

Oral 25 – Process Intensification and Integration

9:30 am (Private Dining, Wednesday)

Oral 127 - Thermally and Chemically Stable Metallic Membrane Reactors for H_2 Generation with CO_2 Capture

Hani A.E. Hawa (Precision Combustion, Inc.)*; Kyle Hawley (Precision Combustion, Inc.); Damian Agi (University of Notre Dame); Alexander Dowling (University of Notre Dame)

Precision Combustion, Inc. (PCI) is developing a novel membrane reactor (MR) for high efficiency and cost-effective H2 generation and separation with simultaneous CO2 capture, suitable for integration with modular, fuel-flexible gasifiers and reformers.

A proprietary Pd-alloy with tailored composition was identified and demonstrated to enhance the thermal and chemical stabilities of the membrane when operated with simulated water-gas-shift (WGS) conditions. An ultra-thin (<5 μm), defect-free film was deposited onto porous ceramic supports for maximizing flux. Unlike performance reported for prior field demonstrations, the membrane demonstrated high H2 permeance and recovery of up to 90% over 1,200 hours of testing with simulated WGS conditions. The membrane was subjected to multi thermal and pressure cycles and maintained a H2 purity of >99.9%, suggesting no leak growth or H2 embrittlement after cycling. The observed stability of the membrane was correlated to the membrane composition and improved microstructure/synthesis.

Using the Department of Energy's (DOE's) Institute of Design and Advanced Energy Systems (IDAES) resources, we validated a WGS-MR model in Pyomo. By combining the Pyomo model with the H2A model developed by the National Energy Renewable Laboratory (NREL), we performed a preliminary design optimization and technoeconomic analysis (TEA), finding our WGS-MR improved the conversion of feed to H2 by ~22%, resulting in ~14% reduction in levelized cost of H2 (LCOH). We then quantified the impact of the enhanced membrane material properties and identified process integration opportunities to further reduce the LCOH to ~$2/kg H2 (with CO2 capture), ~24% reduction in LCOH vs. the baseline cost.

In future work, PCI will mature the WGS-MR to confirm technical and economic viability of producing high purity H2 from gasified biomass/coal blends. We will also address technology commercialization barriers to transition the technology to fielded applications.

10:00 am (Private Dining, Wednesday)
Oral 128 - Specific Energy Analysis of Using Fertilizer-Based Liquid Desiccants to Dehumidify Indoor Plant Environments
Sarah Moussaddy (Oakland University)*; Jonathan Maisonneuve (Oakland University); Sandeep Aryal (Oakland University)

Increasing food production to sustainably feed a growing population requires innovation to improve the efficiency of energy, water, and fertilizer use in agriculture. In this study, we introduce the novel concept of using concentrated fertilizer as a liquid desiccant to efficiently dehumidify indoor plant environments and recycle water vapor for irrigation. This approach closes the water cycle, and eliminates the need for energy-intensive desiccant regeneration that is characteristic of most liquid desiccant systems. The first-ever experimental demonstration of the concept is provided from laboratory testing with a membrane-based dehumidification process using polydimethylsiloxane hollow fibers. Dehumidification is observed for a variety of common fertilizers and across a range of typical indoor plant conditions. Next, a theoretical modeling and parametric analysis is used to examine the impact of key operating conditions on the specific energy of dehumidification of the system. Specific energy is defined here as the ratio of mass vapor removal relative to the liquid desiccant cooling load. Results indicate that specific energy as low as 0.15 kWh/kg is possible under optimized conditions, comparing very favorably with existing dehumidification technologies as well as with recent standards for greenhouse energy efficiency. Future work is proposed to achieve the potential of energy savings and environmentally-friendly plant cultivation with fertilizer-based dehumidification.

10:30 am (Private Dining, Wednesday)
Oral 129 - Feasibility of Membrane-adsorption Hybrid Process for Direct Air Capture
Yi-Chen Huang (The Ohio State University)*; Yang Han (The Ohio State University); Winston Ho (OSU)

Direct air capture (DAC) technologies that extract CO2 directly from ambient air could mitigate CO2 emissions from non-stationary sources. However, the level of atmospheric CO2 is only about 418 ppm, which imposes challenges for DAC if a high CO2 purity is required. Recently, amine-containing polymeric membranes with exceptional CO2/N2 selectivity have been developed for CO2 capture. Nevertheless, a high degree of air compression is needed to provide the transmembrane driving force, which makes the application prohibitive. Another approach for DAC is to employ solid sorbents with high CO2 capacity, but the limited stability of the sorbent and the high regeneration energy usually lead to a high capture cost. In this work, we propose a membrane-adsorption hybrid process for DAC, where a less CO2-selective yet robust physisorbent is used to enrich the CO2 to 45%. This desorbed CO2 is then further purified by a highly CO2-selective membrane, where the higher feed CO2 concentration relaxes the compression requirement. Next, techno-economic analyses of the hybrid process are carried out to evaluate the feasibility of the hybrid process for DAC. Owing to the enriched CO2 feed gas, a single-stage membrane system with the CO2 permeance of 1100 GPU (1 GPU = 10–6 cm3(STP) cm–2 s–1 cmHg–1) and the CO2/N2 selectivity of 100 was able to achieve the 95% CO2 purity with 50% CO2 recovery (overall process).

11:00 am (Private Dining, Wednesday)
Oral 130 - Electrospray 3D Printing to Unlock Customized Membrane Performance
Mert Can Hacifazlioglu (University of Connecticut)*; Jeffrey R McCutcheon (University of Connecticut)

Polyamide membranes have for 40 years dominated the membrane-based desalination field. Interfacial polymerization enabled the manufacturing of polyamide reverse osmosis (RO) and nanofiltration (NF) membranes at scale. Remarkably, little has changed in the materials or the manufacturing approaches chosen to fabricate these membranes since their invention in the early 1980s.
An emerging membrane manufacturing method, electrospray printing, has been shown to be able to fabricate both NF and RO membranes. This process uses high strength electric fields to generate atomized sprays of solvents that carry and distribute monomers homogenously directly onto the support where they subsequently react to form polyamide. This method allows for exquisite control of membrane thickness and chemistry.
Thickness control is possible by adjusting the concentration of the monomer and the number of deposited layers of polymer on the substrate. Chemistry can be adjusted by changing the monomer. This work demonstrates the ability to produce membrane of different thickness and chemistry and to use those control levers to customize membrane properties across a wide range of permeance and selectivity. In particular, we use demonstrate that blending monomers results in producing membranes between NF and RO performance. We also report the membrane yield (i.e. the percentage of membrane tested that were "successful") which is highly dependent on the membrane

thickness. We discuss these results in the context of membrane processes that could benefit from the ability to "dial in" membrane properties for improved desalination treatment train performance.

11:30 am (Private Dining, Wednesday)

Oral 131 - Intensified membrane distillation for treating oilfield produced waters from unconventional wells: Energetics analysis and pilot demonstration

Mahdi Malmali (Texas Tech University)*; Md Hasib Al Mahbub (Texas Tech University); I-Min Hsieh (Texas Tech University); Bosong Lin (Texas Tech University)

The development of a suitable treatment process for produced water (PW) recycling and reuse has been prompted due to the increasing disposal costs and environmental concerns. This study evaluates the performance of a membrane-based separation process for the desalination of PW. Membrane distillation (MD), being an excellent candidate to treat hypersaline water, is considered in this work for PW treatment integrated with pretreatment steps and energy recovery from the product to enhance process performance. Prior to MD, electrocoagulation (EC) and microfiltration (MF) are used as primary and secondary pretreatment processes. These pretreatment steps are necessary to eliminate constituents such as suspended solids, oil, and grease. The pretreated water from the integrated EC-MF process was sent to the vacuum MD (VMD) process that recovered water and concentrated the PW. Vapor compression was incorporated in the VMD which improves latent heat recovery. The integrated EC-MF-VMD process demonstrated in this work has the capacity of treating 1000 L of PW per day, with water recovery in the range of 30-70%. With the EC-MF pretreatments, 90% of suspended solids and turbidity, and 60% of total organic carbon (TOC) were removed, which significantly reduced the PW fouling potential in the MD stage. A custom-designed plate-and-frame VMD module was established which enabled unparalleled service flexibility to process intensification and modularization. This intensified VMD module showed distillate flux from 5 to 66 Lm-2h-1 with the commercial membrane. In addition, a significant improvement in energy efficiency in terms of gained output ratio (GOR = 2.7) and energy consumption per unit of water production (248 kWh/m3) was obtained with vapor compression. In this presentation, we will elaborate on the excellent efficacy of the MD process for PW treatment, and its tremendous potential for PW reuse and recycling.

12:00 pm (Private Dining, Wednesday)

Oral 132 - Low concentration carbon dioxide capture using MEEP-based membranes

Birendra Adhikari (Idaho National Laboratory)*; Amit Nilkar (Idaho National Laboratory); Christopher Orme (Idaho National Laboratory); John Klaehn (Idaho National Laboratory); Hyoenseok Lee (Idaho National Laboratory)

The Elevated level of carbon dioxide (CO2) in the air is the main cause of global warming and climate change. CO2 capture has become necessary not only from the air directly, but also from point sources. Point sources have wide range of CO2 concentrations, but the ones with lower concentration (<3% CO2) has been ignored by CO2 producers because of the high cost of enrichment and capture and high environmental impacts making both economically and environmentally unsustainable. Membrane technologies are considered the most viable CO2 capture solution because of the lower energy cost of operation and present low environmental risks. Our group at Idaho National laboratory (INL) has developed poly[bis((2-methoxyethoxy)ethoxy)phosphazene] (MEEP)-based carbon dioxide selective membranes with CO2/N2 selectivity of above 50 and CO2 permeability of above 450 Barrer. The selective layer of these membranes can be as thin as 100 nm and improved mechanical properties by the addition of C18 functionalized nanodiamonds. A spreadsheet-based technoeconomic analysis (TEA) and life cycle assessment (LCA) model was developed and studied techno-economic and life cycle viability of various membrane and non-membrane based processes. It was determined that a MEEP membrane-based triple staged process has a breakeven cost of production of USD 50/tonne for 99.8% CO2 from the feed source of 1% CO2 in nitrogen (N2). It is determined that the best performing Pebax-based membrane has 464% more, cryogenic capture is 60% to 140% more, pressure swing adsorption is 55% to 165% more and chemical absorption is 10% less to 110% more costlier than MEEP-based capture, respectively. It is also determined that these MEEP-based membrane processes emit at least 42% less equivalent CO2 than that of best performing Pebax-based membrane. Additionally, MEEP-based membrane process emits 34-72% less CO2 than all other conventional separation processes.

Oral 23 – Contaminant Removal from Water

9:30 am (Black Warrior, Wednesday)
Oral 133 - Polyamide Reverse Osmosis Membrane Modification with Diamines to Improve Small Neutral Molecule Rejection
Shahriar Habib (The University of Alabama)*; Madison A Wilkins (The University of Alabama); Steven T Weinman (The University of Alabama)

Reverse osmosis (RO) membranes are the gold standard for seawater desalination. Despite their excellent capacity for salt rejection, RO membranes do not reject small neutral molecules (SNMs), such as urea or boric acid, sufficient to produce potable water. Since the interfacial polymerization reaction is rapid and uncontrolled, the polyamide separation layer contains both network and aggregate free volume holes. SNM rejection is mainly driven by size exclusion, thus reducing the free volume will reduce SNM passage through the membrane. Increasing the degree of crosslinking or decreasing the size of free volume holes on RO membranes is an ideal strategy for achieving this. We hypothesize that if the polyamide layer can be modified with a diamine, then the SNM rejection will be increased. Using chemical modification and heat treatment, a method for enhancing SNM rejection was developed for commercial polyamide membranes. The chemical modification used carbodiimide chemistry followed by diamine coupling. The heat treatment was performed at room temperature (~22 °C) and 63 °C. An electrokinetic analyzer, contact angle goniometer, XPS, ATR-FTIR, and SEM were used to characterize the modified membranes. Membrane performance was evaluated for water permeance, NaCl rejection, and SNM rejection using dead-end filtration. According to the results, both heat treatment and amine modification lead to an increase in the rejection of SNMs.

10:00 am (Black Warrior, Wednesday)
Oral 134 - Electrocatalytic laser-induced graphene membrane for micropollutant removal and biofouling control
Swatantra Pratap Singh (IIT Bombay)*; Najmul Barbhuiya (IIT BOMBAY); Kritika Jasrapuria (IIT BOMBAY)

Over the last few decades, there has been an increase in the number of micropollutants found in different water matrix, which includes dyes, pharmaceutical compounds, endocrine disruptors, pesticides, etc. However, conventional treatment plants are not effective in their removal. Hence, advanced hybrid technologies like electroconductive membranes and advanced oxidation processes are needed to address the challenge of micropollutants and emerging contaminants. Here, we have investigated the potential of an electrocatalytic membrane fabricated from a laser-induced graphene method to remove methylene blue, ciprofloxacin, and amoxicillin from water. The effect of various parameters like voltage, pH, and persulfate addition was also investigated on the methylene blue removal performance. The electrochemical and catalytic activation of persulfate, electro-synthesized hydrogen peroxide at the cathode for reactive species generation in combination with electro-repulsion, electro-adsorption, and electrooxidation resulted in enhanced removal of methylene blue. The removal was seen both in permeate as well as feed under constant retentate-permeate recirculation, thus providing a direction for the retentate disposal issue of membrane filtration. The electrocatalytic membrane filtration was also investigated to remove minute concentrations of ciprofloxacin, amoxicillin, and their mixture in synthetic pharmaceutical wastewater, which also showed good removal of these emerging pollutants. Furthermore, the membranes have also been tested for anti-biofouling and have shown the best result compared to polymeric membranes. Overall, electrocatalytic membrane filtration can provide filtered water, nearly free from these pollutants, and help in the retentate disposal of these filtration systems.

10:30 am (Black Warrior, Wednesday)
Oral 135 - Simultaneous Uranium Extraction and Brine Desalination via Adsorptive Pervaporation by A Bifunctional Graphene Oxide Membrane
Guangcheng Wang (National University of Singapore)*; Sui Zhang (National University of Singapore)

The ocean reserves nearly four billion tons of uranium, which potentially provides an almost inexhaustible supply of nuclear energy. However, directly extracting uranium from seawater is economically challenging due to the extremely low concentration (~3.3 μg·L-1) and high salinity background. Seawater desalination today is commonly operated worldwide, producing brine with higher uranium concentrations. Further concentrating the seawater brine may not only enhance the water recovery and reduce the brine disposal to the environment but also make uranium extraction more feasible. Herein, a pervaporation-adsorption platform was developed for simultaneous desalination and uranium harvesting from seawater brine. The membrane was prepared by modifying graphene oxide with polydopamine and ethylenediamine, followed by glutaraldehyde crosslinking to functionalize it with amino and catechol groups for fast water transport and selective uranium coordination. The membrane exhibits fast desalination of 153.3 ± 0.5 kg·m-2·h-1 with 3.5 wt.% NaCl solution as the feed at 70 ℃ and excellent uranium capture properties of 228.6 ± 2.5 mg·m-2 feeding with 4 mg·L-1 U-spiked solution, enabling concentration of 6 wt.% NaCl solution to over 25 wt.% while harvesting > 70% of the uranium from the solution. Notably, the membrane demonstrates high selectivity of uranium over other competing elements. This study aims to provide a strategy for recovering critical elements from oceans with infinite resources.

11:00 am (Black Warrior, Wednesday)
Oral 136 - Simple and low-cost electroactive membranes for ammonia recovery

Xinyi Wang (University of California, Los Angeles)*; Sungju Im (University of California, Los Angeles)

Ammonia is considered a contaminant to be removed from wastewater. However, ammonia is a valuable commodity chemical used as the primary feedstock for fertilizer manufacturing. In this study, two types of carbon fabrics (non-woven and woven) were used and evaluated for membrane fabrication. A low-cost and highly simple electrically conducting gas stripping membrane (ECM) was developed, characterized, and tested for its ability to extract ammonia from synthetic wastewater. The membrane was constructed by simply bonding commercial carbon fiber (CF) to a commercially available polypropylene (PP) membrane (denoted as CF/PP membrane), using epoxy as the binder. The membrane was paired with a counter electrode and used as a cathode in an electrochemical system. When a cathodic potential is applied to the ECM surface, hydroxide ions are produced at the water-ECM interface, which transforms ammonium ions into higher-volatility ammonia that is stripped across the hydrophobic membrane material using an acid-stripping solution. The membrane exhibited very high electrical conductivity and excellent performance in terms of ammonia recovery, with higher fluxes (an ammonia flux of 141.3 ± 14.0 g.cm-2.day-1) and low energy consumption values (69.2 ± 5.3 kg NH3-N/kWh). The effects of membrane fabrication methods, current density, feed solution pH, and trap solution circulation rate on ammonia flux, membrane stability, and energy consumption were evaluated. It was found that the ammonia flux was sensitive to the current density and acid circulation rate. The simplicity and low cost of the membrane material, coupled with its excellent performance make the direct extraction of ammonia from wastewater a potentially economically attractive possibility.

11:30 am (Black Warrior, Wednesday)
Oral 137 - MIL-100 (Fe) Functionalized Thin-film Nanocomposite Membranes for Enhanced Removal of Uncharged Urea from Water

Tin Le (The University of Alabama)*; Milad Rabbani Esfahani (The University of Alabama)

Urea is widely used in fertilizer production for agricultural purposes which risks runoff into soil and water sources . This poses a concern due to the algal blooms that may form which can poison the environment and cause food source contamination. Thin-film composite (TFC) membranes are among the best performing membranes for water filtration; however, they do not show acceptable rejection for small and uncharged molecules such as urea. Adsorptive membranes combine the power of membrane filtration with the adsorption process and provide enhanced selective removal specifically for small and uncharged molecules. Metal-organic frameworks (MOF) are a new generation of effective adsorbents with excellent tuneability in structure and compatibility with polymeric membranes. We synthesized MIL-100 (Fe), an iron-based MOF, using a green low temperature method, and tested for urea adsorption. The urea adsorption capacity of MIL-100 (Fe) was tested under varying experimental conditions such as pH, temperature, MOF concentration, and urea concentration. The results revealed the superfast adsorption (reached almost 90% removal within 2 min) of neutrally charged urea molecules with a high adsorption capacity. MIL-100 (Fe) was determined to possess a BET surface area of 1174 m2/g and a urea adsorption capacity of 3321 mg/g. In this project, we immobilized the MIL-100 (Fe) on the surface of the polyamide layer of thin-film composite membranes via covalent binding for the fabrication of adsorptive membranes for enhanced removal of urea from water. The MIL-100 (Fe)-TFC membrane showed enhanced permability (7.29 vs 5.52 LMH) and no decrease (47%) in Mg salt rejection, indicating successful grafting without damaging the PA layer. In addition, although both bare TFC and Fe-TFC showed similar rejection of 62% of urea, the MIL-100(Fe)@TFC might have adsorbed a portion of the urea in retentate which is under investigation currently.

12:00 pm (Black Warrior, Wednesday)
Oral 138 - Microcystin-LR destruction and removal from water via enzymatic biodegradation and ultrafiltration

Abelline K Fionah (University of Kentucky)*; Isabel Escobar (University of Kentucky)

Current global outbreaks of cyanobacterial blooms are a cause for concern due to the emergence of their toxic byproducts. Microcystine-LR(MC_LR), a heptatoxin is considered the most toxic of the byproducts. This is because it bioaccumulates and biomagnifies in aquatic environments leading to adverse side effects in higher order organisms. It has been associated with inhibition of protein phosphatases, disruption of formation of the cytoskeletons, promotion of oxidative stress, liver tumors leading to liver cancers. Cleavage of the MC-LR toxin at various bond has been proven be an effective degradation method for this toxin. However, the position of the Leucine and arginine amino acids makes it hard to degrade the toxin. Protein Trads containing MlrA, MlrB, and MlrC have been proven to effectively cleave the MC-LR at the different bonds therefore deactivating the active sites of this toxin. Literature shows that biotoxins can be removed from aquatic environments via membrane ultrafiltration. This project focuses on evaluating the effectiveness of an ultrafiltration membrane in removing the MC-LR degradation byproducts from water. A sulfone functionalized poly-ether ether ketone(S-PEEK) blended with polysulfone (PSf) was utilized. Full characterization of the membrane as well as the enzymatic degradation by products was carried out using SEM, XPS, FT-IR, TOC measurements, etc. the results from this project analyzed via MS-LC showed that the MlrA was capable of linearizing MC-LR toxin and that the S-PEEK-PSf/NMP membrane can remove the degradation byproducts of this toxin from water. The results indicated that coupling the two systems has the potential to not only facilitate the recyclability of the enzyme, but also enhance enzyme properties such as viability and stability.

Oral 24 – Membrane Fouling

9:30 am (Capstone, Wednesday)

Oral 139 - Energy reduction in ceramic microfiltration using supersaturated carbon-dioxide-enhanced backwash

Weiming Qi (Clemson University)*; David Ladner (Clemson)

Membrane fouling is the key difficulty during the filtration process. Frequent backwash and chemical-enhanced backwash (CEB) are applied periodically between filtrations to remove reversible fouling and irreversible fouling, respectively. However, the limitations of backwash and CEB are obvious that backwash is not robust enough to recover other types of fouling than cake layers, while CEB has a better efficiency than backwash; however, the chemical cost, cleaning residual, and membrane degradation are problematic. To avoid these drawbacks, supersaturated CO2 solution has been tested to backwash the ceramic membrane. Unlike air-assisted backwash, which uses air bubbles scouring the feed side surface, supersaturated CO2-enhanced backwash (SCEB) forces CO2 solution through membrane pores as normal backwash. The CO2 nucleation happens in the membrane pores and on the membrane surface to help enhance backwash efficiency. Three membranes with different serving ages were tested with filtration of lake water, activated sludge, and fat, oil, and grease (FOG) wastewater, which contributed to a total of nine experiments. In each experiment, seven filtrations were separated by six cleaning steps. The first, third, and fifth cleaning steps were DI water backwash; the second, fourth, and sixth cleaning steps were SCEB. By using this approach, DI water backwash was directly compared with SCEB one after another. The results show that SCEB saved from 0.7% to 17% of the membrane filtration energy. FOG wastewater benefitted from SCEB more than lake water and activated sludge. The results also suggest that the newer membrane showed more energy saving from SCEB.

10:00 am (Capstone, Wednesday)

Oral 140 - The role of surfactants in membrane fouling

Naama Segev-Mark (Technion)*; Kirti Sankhala (Technion-Israel Institute of Technology); David Jassby (UCLA); Eric Hoek (UCLA); Guy Z Ramon (Technion - Israel Institute of Technology)

Surfactants are common additives in chemical, biomedical and biotechnological processes, thus are likely to be found in wastewater, with a particular case being oil/water emulsions. Membrane separation can successfully separate these emulsions but can suffer from severe fouling, normally attributed to membrane surface coverage by oil droplets, which causes flux decline. The role of surfactants in the observed fouling is far less understood and is the focus of the present study. By means of direct confocal microscopic observation, we assess the correlation between hydrodynamics, surface properties and solution composition, linking the microscopic events observable through the microscope, with macroscopic system behavior via pressure-flux relations. In particular, we examine how different membrane surfaces react to different surfactants.
Our results correlate membrane performance, i.e. flux losses, with real time 3D images of droplets at the surface of a 100kDa PES UF membrane. When a membrane is exposed to non-ionic surfactant (TX-100), the system loses 70% of its performance, while the surface is covered with only a few small (<20 μm) droplets. In contrast, filtering oil

alone results in only 20% flux decline, even after longer times and with far greater fouled surface coverage. Extending these results to other polymeric membranal surfaces (PAN and PVDF) with similar MWCO and different surfactants (non-ionic and ionic) suggest that indeed the surfactants are major contributors to the observed permeate flux decline. These results motivates re-consideration of fundamental questions regarding the fouling mechanisms of emulsions, and their implications for fouling mitigation.

10:30 am (Capstone, Wednesday)

Oral 141 - In-situ measurements to differentiate homogeneous and heterogeneous nucleation in membrane distillation: effects of spacer geometry and antiscalants

Mahdi Malmali (Texas Tech University)*; Hadis Matinpour (University of California - Santa Barbara); I-Min Hsieh (Texas Tech University); Bosong Lin (Texas Tech University)

Fouling and scaling in membrane distillation (MD) has remained a significant challenge that leads to reduced membrane throughput and possibly membrane wetting. Prior reports have elucidated the scaling behavior based on indirect measurements (flux) or ex-situ analyses (e.g., SEM and EDX). Therefore, these findings are limited to conclusions that do not distinguish the contributions from homogeneous and heterogeneous nucleation. In this presentation, a non-invasive, in-situ, and real-time imaging setup will be discussed that was used to study the scaling mechanism. This device assisted with distinguishing different phases of scaling during the filtration tests. The scaling mechanism of a single-component system with sodium chloride, strontium sulfate, calcium sulfate, and calcium carbonate in vacuum MD (VMD) was investigated. First, the effect of flow hydrodynamics on gypsum scaling and membrane wetting was evaluated. Then, the effect of the natural organic matter (NOM) on gypsum scaling was systematically investigated and compared against the model test.
It was found that concentration polarization (CP) was mitigated by improved hydrodynamics, which played an important role in heterogeneous nucleation and crystal growth; the crystal deposition on the membrane surface from bulk phase crystallization was affected by the hydrodynamics. Humic acid (HA) promoted gypsum heterogeneous nucleation due to HA/calcium ions interaction. The mitigation of crystal growth on gypsum scaling by HA was observed and attributed to the adsorption of HA on salt crystals. Higher organic fouling expedited the surface crystallization resulting from heterogeneous crystallization and enhanced the mitigation of crystal growth. Sodium alginate (SA) acted as seeding material for crystallization and enhanced the nucleation of gypsum scaling was observed in the bulk phase crystallization. Polycarboxylate antiscalant at 5 ppm delayed the onset of nucleation in the bulk phase.

11:00 am (Capstone, Wednesday)

Oral 142 - Antimicrobial Properties of Chitosan-Graphene Oxide Composites

Christopher S Griggs (U.S. Army ERDC)*; Audie Thompson (U.S. Army ERDC); Justin Puhnaty (U.S. Army ERDC)

Graphene oxide (GO) is a promising candidate material for antimicrobial membranes due to oxidative debris (OD) that can serve as reactive oxygen species (ROS) generation sites. However, the underlying mechanisms of oxidation have not been resolved. This study demonstrates the role of the superoxide radical (O2- •) in the oxidative properties of GO, coupled with the electrostatic interactions of cationic chitosan (CS) in GO composite membranes (CSGO). Biofilm studies confirm growth inhibition is attributed to both contact-based physical interactions coupled with oxidative stress in a "trap and treat" mechanism. Bacteria applied to the surface of the CSGO composite were unable to thrive on the material, thus indicating that its surface was an unfavorable environment for cell growth instigating interruptions in vital cellular processes due to stress on the membrane from interactions with GO materials. Both superoxide dismutase (SOD) assay and electron spin resonance (ESR) confirmed the generation of the ROS, superoxide. The oxidation of glutathione (GSH) confirms the presence of superoxide at levels capable of inducing oxidative stress. The synergistic mechanisms of CSGO produce a viable antimicrobial platform for a broad spectrum of pathogens present in environmental and wastewater applications. For water treatment, production of reactive oxygen species originating from GO containing OD with oxygen based functional groups from the CSGO composite can further discourage biofouling from bacteria overgrowth. Due to scalability and performance of the CSGO composite for both bacterial entrapment and cellular inhibition make this material an attractive option as an antifouling membrane in water treatment applications. The results demonstrate that the composites fabricated in this study can be utilized in the future to inactivate pathogens in the medical, food processing, and water treatment industries as well as in the field of environmental remediation.

11:30 am (Capstone, Wednesday)

Oral 143 - Spacers for Mass Transfer Enhancement in Flat Sheet Membrane Modules – An Upper Bound?

Arsalan Sepehri (University of Toledo); Glenn Lipscomb (University of Toledo)*

Spacers are used to create flow channels and mix fluid within the channel to enhance mass transfer and reduce fouling potential. Unfortunately, spacers increase pressure drop. The increase in mass transfer coefficient and pressure drop depend strongly on spacer geometry.

The literature that addresses spacer performance is reviewed. Performance results often are compared to identify spacers designs with the most desirable characteristics (combination of mass transfer coefficient and pressure drop). These comparisons are confounded by differences in membrane transport and the boundary conditions that couple mass fluxes between the contacting fluid and membrane domains. Additionally, most studies do not examine the effect of flow channel length on mass transfer. Boundary layer resistance typically increases with flow channel length because fluid adjacent to the membrane is not completely mixed with the bulk as fluid flows through the spacer.

An apparent upper bound emerges upon examining the available literature – for a given pressure drop, the mass transfer coefficient cannot be increased above a maximum value. The upper bound is related to the trade-off between increased mass transfer performance and increased pressure drop. Increases in mass transfer coefficient typically are achieved by increasing fluid shear rate adjacent to the membrane surface. Increased shear rates lead to increased pressure drops and the emergence of the apparent upper bound.

The location of the upper bound is consistent with effectively remixing the fluid in the concentration boundary layer every ~100 microns. The factors that control this effective remixing length are explored through a simple model of spacer performance based on alternating accelerating (high shear and high mass transfer rate) and decelerating (low shear and low mass transfer rate) flows along the flow path.

12:00 pm (Capstone, Wednesday)
Oral 144 - Real-Time Induced Magnetic Vibrational Based Antifouling Mechanism for Ultrafiltration (UF) membrane

Jasneet K Pala (The University of Alabama)*; Ryan Tracy (The University of Alabama); Milad Rabbani Esfahani (The University of Alabama); Nima Mahmoodi (The University of Alabama)

Membrane fouling is still one of the main limiting factors, even though membrane technologies have been established across the world as one of the most efficient water treatment. Fouling reduces membrane separation efficiency by reducing membrane permeability and rejection, shortening membrane lifespan, and increasing process costs due to higher washing requirements. Numerous works have been published on the chemical modification of membranes to reduce fouling. However, several negative side-effects, such as reduced flux or selectivity, have been reported due to these approaches. In this work, we report the implication of real-time induced magnetic vibration as a green method for membrane antifouling without the need to affect the permeability and selectivity of the membrane. The attachment of foulant particles on the membrane surface will affect the fouling initiation mechanism and, consequently, the membrane fouling. The implication of magnetic vibration on the membrane surface will prevent or delay the deposition of particles on the membrane surface, resulting in antifouling or lower-intensity fouling. The results show that the deposition of foulants (humic acid and a model particle) on the membrane surface changed by applying different frequencies. The two-dimensional wave equation was used to identify the effect of frequency on the deposition of particles on the membrane. The experimental investigation confirmed the modeling data where at a higher frequency of 311 Hz the foulant formed a circular pattern near the membrane edges, and at a lower frequency, 190 Hz, the foulants formed a circle pattern in the center of the membrane. A cross-flow filtration system was used to examine the effect of real-time magnetic vibration on the humic acid fouling on the UF polysulfone membrane. The result shows that applying magnetic vibration (28W) reduced the fouling intensity by almost 40 % compared to membranes tested without the magnetic field.

Oral 22 – Membranes for Gas Separations

9:30 am (Yellow Hammer, Wednesday)
Oral 145 - Co^{2+}-based supramolecular networks achieving extraordinary CO_2/O_2 selectivity

Taliehsadat Alebrahim (The State University of New York at Buffalo)*; Narjes Esmaeili (The State University of New York at Buffalo); Gengyi Zhang (The State University of New York at Buffalo); Thien N Tran (University at Buffalo, The State University of New York); Haiqing Lin (The State University of New York at Buffalo)

Co2+ ions can reversibly and selectively interact with O2 and thus are expected to influence O2 transport properties. However, Co2+ ions are difficult to dissociate in polymers, and thus, very few studies demonstrate their use for oxygen separation. Herein, we develop Co2+-induced supramolecular networks based on amorphous poly(ethylene oxide) (XLPEO) with superior CO_2/O_2 separation properties. Specifically, supramolecular networks were prepared by photopolymerization of poly(ethylene glycol) diacrylate (PEGDA), poly(ethylene glycol) methyl ether acrylate (PEGMEA), and different salts such as $Co(ClO_4)_2$ and $Co(BF_4)_2$. Co^{2+} ions conjugate with the ethylene oxide in XLPEO, as validated by FTIR, XRD, and DSC. For example, adding 17 wt% $Co(BF_4)_2$ in XLPEO increases the Tg from -63 to -27 ºC, O_2 solubility by 60 times from 0.18 to 11 cm^3(STP)/(cm^3 atm), and O_2/CO_2 solubility selectivity by 10 times from 0.11 to 11 at 35°C. Surprisingly, the presence of Co^{2+} dramatically reduces O_2 permeability because of the retarded O_2 diffusion. For example, addition of 17 wt% of $Co(BF_4)_2$ decreases O_2/N_2 permselectivity from 2.6 to 0.6 and increases CO_2/O_2 permselectivity from 20 to 60. The supramolecular structure containing 2 wt% $Co(BF_4)_2$ shows CO_2 permeability of 109 Barrer and CO_2/O_2 selectivity of 50, surpassing Robeson's upper bound. Thin-film composite membranes based on the supramolecular networks were successfully prepared and exhibit pure-gas CO_2 permeance of 770 GPU and CO_2/O_2 selectivity of 77 at 35 oC, which is very attractive for post-combustion carbon capture or controlled atmosphere for food storage. We will also discuss the incorporation of Co^{2+}-based metal-organic frameworks (such as ZIF-67) in XLPEO and glassy polymers on O_2/gas separation performance. This work unravels an interesting platform of metal ions-based supramolecular networks with unique gas sorption behaviors and exciting gas separation properties.

10:00 am (Yellow Hammer, Wednesday)

Oral 146 - Crown Ether-Based Polyimides for Sour Gas Separation

Dana Wong (Aramco Americas)*; Elizabeth Haddad (Aramco Americas); John Yang (Aramco Americas); Patrick Wright (Aramco Americas)

As the one of the world's primary energy sources, natural gas accounts for a quarter of global energy consumption. However, many natural gas reservoirs contain toxic contaminants such as sour gases (e.g., CO_2 and H_2S) which must be removed before combustion to meet industry standards. Crown ethers are a flexible and bulky monomer possessing high CO_2 affinity due to their abundance of oxygen-containing ether groups. 6FDA-DAM is a rigid polymeric structure containing $-C(CF_3)_2$ groups from the anhydride and methyl groups from the diamine which increase polymeric chain spacing and the free fractional volume, resulting in increasing membrane permeability. To achieve higher separation performance (e.g., H_2S/CH_4 and CO_2/CH_4 selectivities) for sour gas separation, many 6FDA-DAM based copolyimides have been synthesized by employing the rigid aromatic amines as comonomers. The incorporation of the crown ether can help increase membrane separation performance. In this presentation, two crown ether-based copolyimide membrane materials were synthesized by comprising 18-crown-6 ether and 24-crown-8 as comonomers in the polymer backbone. The synthesized materials were characterized by FTIR, NMR, TGA and DSC. The sour gas separation performance of these membranes was examined under industrially relevant feed streams and testing conditions. The obtained co-polyimide membranes displayed enhanced (H_2S and CO_2)/CH_4 mixed gas selectivity, at high H_2S content of 20% and feed pressure up to 800 psi.

10:30 am (Yellow Hammer, Wednesday)

Oral 147 - Modeling Gas Permeation in Amorphous Polymer Membranes Using Non-Equilibrium Molecular Dynamics Simulations

Mohammed N Al Otmi (University of Florida)*; Janani Sampath (University of Florida)

Advances in material synthesis and membrane fabrication techniques have given rise to many polymer candidates for separation applications. Conducting a thorough screening of these materials using experimental techniques can pose a significant challenge, necessitating a considerable investment of time and resources. While equilibrium molecular dynamics (EMD) simulations provide novel molecular insights about a membrane's performance under realistic process conditions, they require lengthy and expensive simulations. Non-equilibrium molecular dynamics (NEMD) simulations can potentially circumvent such extensive calculations and provide insights into the underlying transport mechanisms, such as the penetrant/membrane interactions, the effect of pressure and temperature on transport, and the role of the membrane structure rearrangement. This study uses NEMD to predict the permeability of three different glassy polymer structures: polystyrene (PS), polymer of intrinsic microporosity (PIM-1), HAB-6FDA thermally rearranged polymer (TRP) with hydrogen, ethane, and ethylene. We apply the concentration gradient (CG) method in which we impose an external force potential on a small part of the system establishing a concentration gradient and moving molecules across the membrane. NEMD simulations allow for a direct calculation of permeability using the flux data and pressure difference. By analyzing the concentration profile and penetrant average velocity, we can visualize the transport channels and the penetrant motion across the membrane. By increasing the applied force, we examine permeation under different pressures and illustrate how the chain rearrangement impacts transport. Chain rearrangement is assessed by examining the translational and orientational dynamics of different segments in the polymer. NEMD simulations can be a powerful tool for the high-throughput screening of new materials, allowing us to explore the material design space more efficiently.

11:00 am (Yellow Hammer, Wednesday)

Oral 148 - Selective H_2S/CO_2 Separation using Sterically Hindered Amine Membranes

Shraavya Rao (Ohio State University)*; Xuepeng Deng (The Ohio State University); Yang Han (The Ohio State University); Li-Chiang Lin (The Ohio State University); Winston Ho (OSU)

H2S, a highly toxic and corrosive gas, is a common contaminant in syngas and natural gas. To minimize occupational hazards, it is necessary to selectively remove H2S from such CO2-containing fuel gases before further processing. Consequently, there is a need to develop technologies for selective H2S/CO2 separation. This work describes the development of amine-based facilitated transport membranes (FTMs) with high H2S/CO2 selectivity.

11:30 am (Yellow Hammer, Wednesday)

Oral 149 - Sweetening Sour Gas with Hybrid Membranes

Patrick T Wright (Aramco Americas)*; John Yang (Aramco Americas)

Natural gas accounts for more than 20% of the world's energy consumption and is responsible for almost a quarter of global electricity production. Raw natural gas is often contaminated with high concentrations of hydrogen sulfide (H2S) and carbon dioxide (CO2). Such contaminated natural gas is commonly referred to as sour gas. More than 40% of currently available global natural gas resources are sour. Indeed, the Middle East region is known for having sour gas, with H2S concentrations reaching up to 30 mol.% along with significant amounts of CO2. Due to the acidic nature of H2S and CO2, sour gas is highly corrosive when moisture is present. H2S is also highly toxic and poses a significant safety risk. Thus, before sour gas can be transported through a pipeline, it must be "sweetened" by removal of its acidic impurities. Applying membrane-based separation techniques instead of amine adsorption can help reduce the high cost of separating sour gas, but there are currently no membranes available that can simultaneously remove both H2S and CO2 from industrially relevant feed streams under realistic testing conditions. In this presentation, we will introduce our recent progress on the simultaneous removal of H2S and CO2 from highly sour natural gas – a mixture with high concentrations of both H2S and CO2 – using mixed matrix membranes. While the strategy of incorporating fillers into the polymer membrane matrix to form mixed matrix membranes (MMMs) has been proposed for several decades, this is a unique opportunity for MMMs to succeed in natural gas sweetening by maximizing the advantages of both polymer and fillers. Our work suggests a new direction for the design of membrane materials that is focused on real-world industrial gas processing.

12:00 pm (Yellow Hammer, Wednesday)

Oral 150 - Identifying key factors to unlock plasticization resistance in microporous polymer membranes

Katherine Mizrahi Rodriguez (MIT); Sharon Lin (MIT); Albert Wu (MIT); Kayla Storme (MIT); Taigyu Joo (Massachusetts Institute of Technology); Aristotle F Grosz (MIT)*; Naksha Roy (MIT); Duha Syar (MIT); Francesco Benedetti (MIT); Zachary Smith (MIT)

As the need for robust gas separation membranes accelerates at an unprecedented rate, there exist core challenges that must be overcome before reaching broader and cutting-edge industrial applications. One of the most prominent obstacles is plasticization, a phenomenon in which the sorption of condensable gases into the polymer increases chain mobility. Above the "plasticization pressure," this behavior is particularly detrimental to a separation process as it increases the permeability of all penetrants, which can lower the overall permselectivity significantly. Plasticization arises as an issue inherent to many critical contemporary systems (notably, CO2- and H2S-based separations), especially for microporous polymers that otherwise have outstanding performance.

In order to identify key material properties that grant plasticization resistance, we have conducted an exhaustive review of microporous polymer literature spanning from the advent of PIM-1 in 2004 to present day, compiling the results of over 40 unique polymers to compare plasticization pressures and, where possible, to compare pure- and mixed-gas experiments side-by-side. Moreover, we describe the theory behind plasticization, including several gas transport models which address plasticization behavior. Further, we discuss several promising synthetic/material approaches researchers have leveraged in an effort to mitigate plasticization, such as tailoring backbone chemistry to induce chain rigidity or hydrogen bonding, post-synthetic modification via repeat unit rearrangement or crosslinking, or using multi-component systems such as copolymers, blends, or mixed-matrix membranes. Additionally, we compare design strategies with sorption isotherms and plasticization behavior to decouple the role of sorption and polymer structure from plasticization behavior. We conclude by recommending certain concepts with literature precedent that show the largest potential influence on controlling plasticization.

Abstracts – Oral Presentations, Wednesday

Presenting authors are indicated by an asterisk*.

Oral 26 – High Salinity Streams and Electrochemical Separations 3

2:00 pm (Druid, Wednesday)
Oral 151 - Sulfonated Polystyrene Membranes Near the Percolation Threshold for Hypersaline Electrodialysis Desalination

Yuxuan Huang (Columbia University)*; Marshall Tekell (Columbia University); Hongxu Chen (Columbia University); Sanat Kumar (Columbia University); Ngai Yin Yip (Columbia University)

The sustainable management of hypersaline brines (typically ≥70,000 ppm TDS) has become a pressing demand. Although desalination offers an attractive option, energy-efficient hypersaline desalination is still technically challenging. As a mature technology widely applied in brackish water desalination, electrodialysis, ED, has obtained increasing attention regarding hypersaline applications for its unique advantages. However, as the key component in ED, current ion-exchange membranes (IEMs) are unsuitable for hypersaline operations due to 1) diminished permselectivity in high ionic strength environment and 2) water transport resulting from transmembrane osmotic pressure difference. In this study, we develop sulfonated polystyrene membranes that can satisfy the two critical needs. Polystyrene-r-sulfonated polystyrene copolymers with carefully controlled sulfonation levels were neutralized with sodium hydroxide and then fabricated into thin films through solvent evaporation. Wide-angle X-ray scattering analysis and measurements of ion exchange capacity and conductivity indicate a critical ion cluster percolation threshold for ions to permeate through the membrane. Near the percolation threshold, fabricated membranes show permselectivity ≥0.82 when characterized by the static method with 4.0 and 0.8 M NaCl solutions, which is superior to commercial IEMs (significantly diminished permselectivity of ≈0.76). Furthermore, when challenged in the ED process with 4 M NaCl solution, excellent permselectivity of ≥0.96 is observed (commercial IEM is ≈0.74). Additionally, sulfonated polystyrene membranes also exhibit water permeability as low as ≈60 mL μm/(m2 h bar), with commercial IEMs typically between 100 to 1000 mL μm/(m2 h bar). The study demonstrates a facile approach to fabricating IEMs capable of maintaining high permselectivity but low water permeability under high solution concentrations, thus enabling hypersaline ED desalination and other electromembrane processes.

2:30 pm (Druid, Wednesday)
Oral 152 - Controlled Localized Metal Organic Framework Synthesis on Anion Exchange Membranes

Harm Wiegerinck (University of Twente)*

In electrodialysis (ED) of monovalent and multivalent ion mixtures, the ion-exchange membranes (IEMs) become comparatively more enriched in counterions of the higher valency. This can result in higher fluxes of these multivalent ions. Coating an IEM with a layer that selectively lowers the transport rate of the multivalent ions makes it possible to selectively remove the monovalent ions. For example, this allows removal of chloride from phosphate and sulfate in greenhouses at potentially lower electrical energy costs.
Metal-Organic Frameworks (MOFs) are a promising coating material for IEMs, since it has been shown in literature that thin MOF films exhibit preferential transport of monovalent ions over multivalent ions. This selectivity is caused by the angstrom-sized pores in the MOF cage, leading to a dielectric exclusion/dehydration-based selectivity. Multivalent ions generally have larger hydration radii and require more energy to dehydrate; therefore monovalent ions can pass the MOF easier compared to multivalent ions.
MOF modified IEMs have already been demonstrated in open literature, e.g., Xia et al. (2022). As an alternative method, we take advantage of the properties IEMs to locally control the concentrations of both metal and organic linker. Our method is a modified version of the contra-diffusion synthesis of Yao et al. (2011), where the metal ion and organic ligand are now separated by an AEM. Since the metal ion is locally depleted by Donnan exclusion and its transport is hindered through the AEM, this resulted in the formation of a several micrometers thick MOF film exclusively on the metal solution side. Furthermore, this method allows one to control the MOF crystallization process by tuning the precursor concentrations. Our work demonstrates the advantage of using the properties of AEMs to locally control MOF film synthesis and opens up opportunities to coat or pattern ion-exchange materials for use in ED.

3:00 pm (Druid, Wednesday)

Oral 153 - Assessing the Suitability of Nanofiltration for Pretreatment of Lithium-Enriched Brines

Juan Zhai (Texas Tech University)*; Yuexiao Shen (Texas Tech University); Mahdi Malmali (Texas Tech University); rajesh khare (Texas Tech University); Amrika Deonarine (Texas Tech University); Shubhra Bhattacharjee (Texas Tech University); Adegbola Balogun (Texas Tech University)

Geothermal brines from Salton Sea have been identified as a potential domestic source of lithium. The challenge of extracting lithium from the complex brine has attracted lots of interests in advanced strategies such as nanofiltration (NF).This work examines the technical and economic aspects of NF systems using simulated brine with two different concentrations, under real brine concentration (300g/L) as scenario 1 and under diluted real brine concentration (10 times) as scenario 2, respectively, to evaluate if NF is suitable for brine pretreatment to enrich lithium and remove most of the divalent cations. Commercial NF270, DL, and NF90 membranes have been tested. The results showed that both NF270 and DL membranes were permeable under scenario 1. However, the permeability values were extremely low (1Lm-2h-1bar-1).The rejections of Li+, Mg2+, Na+, K+, and Ca2+ for NF270 vs. DL membrane were (-16.8%±0.48vs.-14.3%±1.41), (17.7%±1.42vs.24.5%±1.04), (12.7%±1.12vs.16.8%±0.45), (12.3%±0.63vs.18.2%±0.52), (10.1%±0.46vs.7.7%±1.04), respectively. Scenario 2 showed the permeabilities of NF270 and DL were 4.52 and 2.53 Lm-2h-1bar-1. It can be inferred that the feed salinity impacted NF permeability and rejection properties. With 2g/L of Na+ and Ca2+ solution, the NF270 and DL membranes achieved a rejection of 16.2%vs.24.2% and 50.1%vs.77.6%. However, when the feed concentration was increased to 200g/L, the rejection of both membranes were reduced less than 10%.NF90 has more rigid structures but still suffered from poor Ca2+ rejection. The rejection was not affected at different pH due to high salinity conditions. The computational simulation was employed to explain the mechanisms of NF rejection at different salinity levels and salt transport. Techno-economic analysis was conducted to determine the economic feasibility of NF-based enrichment approaches, compared with another commonly used sorbent technology.

3:30 pm (Druid, Wednesday)

Oral 154 - Ion transport in ion exchange membranes: water confinement effects at low hydration

José Carlos Díaz (University of Michigan)*

Ion exchange membranes (IEMs) have become great candidates for electrochemical water desalination. Low-hydration IEMs possess higher charge densities and superior co-ion/counterion selectivity in comparison to highly hydrated IEMs. IEMs have also found widespread use in energy technologies and resource recovery as selective ionic conductors where counterion/counterion transport selectivity is most important. However, the exact molecular properties of IEMs that influence counterion diffusion selectivity have not been systematically investigated in the literature. Counterion transport in IEMs for electrochemical applications is often quantified by ionic conductivity. The ionic conductivity is significantly affected by the water content and the charge density in IEMs. Water content and charge density are often challenging to control independently. Fundamental studies investigating the influence of water content and ion hydration as an isolated variable on membrane performance can guide the design of novel membranes with increased counterion transport selectivity.

In this study, a set of crosslinked homogenous IEMs with varying water content and constant fixed charge concentration were synthesized via free radical polymerization. The IEMs with varying water content serve as a platform to investigate counterion transport selectivity at different water content. Bulk counterion diffusion coefficients were measured via ionic conductivity. The ion interactions were investigated via FTIR, dynamic mechanical analysis, and activation energy of ion conduction. The internal structures of the IEMs were characterized through small-angle x-ray scattering. The states of water in the IEMs were studied via differential scanning calorimetry and pulse-field gradient NMR. In this presentation, we will discuss the effects of water content, states of water, ion-ion interactions, and ion-polymer interactions on counterion transport selectivity in IEMs.

4:00 pm (Druid, Wednesday)

Oral 155 - Extending the reverse osmosis to osmotic mediation for brine management

Ali Naderi Beni (Purdue University)*; Iraj Ghofrani (Sharif University of Technology); Ali Mousavi (Sharif University of Technology); David M Warsinger (Purdue)

Treating high salinity streams is one of the challenging steps in reaching zero liquid discharge, as conventional methods require high energy consumption. This research introduces a new countercurrent membrane with cascade recycle with multiple system feed inlets. It uses osmotic mediation methods, including split-feed counterflow reverse osmosis (CFRO), split-brine CFRO, and low-salt-rejection reverse osmosis (LSRRO). The standalone performance of the three methods is investigated based on the proportion of concentrate and permeate flow rate determined by the split factor in CFRO methods and salt permeability coefficient in LSRRO, which affect salt and water flux as well as brine and dilute salinities. Also, pinch point analysis is introduced to assess the performance of CFRO methods in reaching high salinities. Furthermore, a techno–economic optimization is performed on the integrated systems implemented with osmotic mediation methods to dewater a feed of 70 g/l salinity to brine with 234 g/l salinity. The results showed that the integrated systems with split-feed CFRO, split-brine CFRO, and LSRRO can perform the process with unit water cost of 3.63, 3.44, and 3.77 $ per cubic meter of product fresh water, respectively. Their corresponding specific energy consumption is 9.47, 8.66, and 16.82 kWh per cubic meter of the product stream. Moreover, the effect of system feed salinity on the system performance was analyzed for the salinity range of 30 to 90 g/l. The results showed that when the salinity is less than 70 g/l, the optimal distribution of the system feed is through the first inlet, while when exceeding this threshold, it should be fed to the second system inlet. This research helps infer future decisions on implementing appropriate membrane properties and suitable methods to achieve energy savings by inferring the abilities of different methods to facilitate the decisions on designing membranes capable of boosting the efficiency of the process.

4:30 pm (Druid, Wednesday)

Oral 156 - Understanding mixed salt partitioning in model poly(ethylene oxide)-based membrane materials

Everett S Zofchak (The University of Texas at Austin)*; Aubrey Quigley (The University of Texas at Austin); Jordyn Yoh (The University of Texas at Austin); Venkat Ganesan (The University of Texas at Austin); Benny Freeman (The University of Texas at Austin)

Recently, there has been significant interest in developing membrane materials for ion-selective separations (e.g., lithium from saline brine). Most ion-selective membrane studies focus on embedding ligands or nanomaterials into polymers to imbue these materials with ion-selective properties. However, these studies typically characterize membrane performance under single salt conditions, which may not accurately reflect the behavior observed in complex salt mixtures. Further, when mixed salt studies are present, they are typically performed at a single ionic strength or salt composition, obscuring the fundamental origins of differences between the mixed and single salt case. In this study, we probe the influence of solution ionic strength and salt mole fraction on salt partitioning from binary mixtures of alkali chloride (e.g., LiCl/NaCl/, LiCl/KCl, and NaCl/KCl) salts in a model poly(ethylene glycol) diacrylate membrane. At a constant concentration of a given salt, we observe that salt's partitioning to be substantially different in the pure and mixed environments due to large differences in solution phase thermodynamics (e.g., mean-ionic activity). We derive a thermodynamic model to describe the partitioning of individual ions from salt mixtures, where the parameters for the model are computed from molecular dynamics simulations. We see qualitative, and in some cases semi-quantitative, agreement between our model and the experimental results, indicating the predictive power of our modeling approach. Our results provide a sound theoretical framework for understanding mixed salt partitioning phenomena, while simultaneously motivating the need for rigorous, standardized procedures for reporting ion-ion selectivity in membrane materials.

Oral 27 – Water Reuse

2:00 pm (Yellow Hammer, Wednesday)

Oral 157 - Advancements in Water Treatment: Concentrating RO Residuals Contaminated with PFAS through Closed Circuit RO and Membrane Distillation

Tae Lee (US EPA)*; Mallikarjuna Nadagouda (Wright State University); Hyunsik Kim (University of Cincinnati)

Reverse osmosis (RO) has demonstrated remarkable capability in removing per- and poly-fluoroalkyl substances (PFAS) from water. However, managing PFAS-rich RO reject water residuals may be a major challenge in the future due to their adverse effects on the environment and human health, and what this may mean in terms of future discharge permit requirements along with any concerns regarding potential liability. Development of effective PFAS treatment techniques for RO brine streams is an important area of research to assure that public water systems can continue to maintain discharge permits in a changing regulatory environment, and to maintain peace of mind regarding avoiding future disputes with downstream communities. In this study, we evaluated closed-circuit RO (CCRO) and membrane distillation (MD) as methods to further reduce the volume of the RO residual with the aim of reducing the compliance costs and improving the efficacy of viable residual treatment options such as advanced oxidation, crystallization, and granular activated carbon (GAC). CCRO enables high-recovery operation by recirculating the RO concentrate, while continuously recovering clean permeate water through pressure-driven membrane processes. A bench-scale RO-GAC hybrid system was constructed to demonstrate CCRO's applicability for treating RO residuals. MD is useful for treating difficult-to-treat water sources since the process is thermally driven and not limited by high osmotic pressure or concentration polarization. However, studies have found that the presence of PFAS in the feed water may cause premature membrane fouling and pore wetting due to the rapid surface diffusion of PFAS. We evaluated the benefits of a hydrophilic coating using polyvinyl alcohol (PVA) crosslinked with sulfosuccinic acid (SSA) on a commercial polytetrafluoroethylene (PTFE) membrane to impart an anti-wetting property by limiting the hydrophobic interaction of PFAS's hydrophobic tails with MD membranes.

2:30 pm (Yellow Hammer, Wednesday)

Oral 158 - Backwashable hollow-fibre membranes enable ultrahigh recovery nanofiltration process for effluent reuse

Chidiebere S Nnebuo (Department of Desalination and Water Treatment, Zuckerberg Institute for Water Research, The Jacob Blaustein Institutes for Desert Research, Ben-Gurion University of the Negev, Beersheba 84990, Israel)*; Oded Nir (Department of Desalination and Water Treatment, Zuckerberg Institute for Water Research, The Jacob Blaustein Institutes for Desert Research, Ben-Gurion University of the Negev, Beersheba 84990, Israel)

Climate change escalation and worsening global water scarcity decrease the available freshwater supply. Hence, the need for alternative freshwater sources. Nanofiltration (NF) membrane has emerged as a promising effluent treatment technology - an alternative freshwater source. However, membrane fouling limits NF wastewater effluent treatment to 70-85% water recovery, generating large volumes of retentate stream. The high cost of retentate management often limits wastewater effluent desalination. To tackle this challenge, we tested the application of physical backwashing of hollow fibre polyelectrolyte multilayer (HF-PEMM) NF membrane as a strategy to achieve high water recovery (≥95%) in real wastewater effluent treatment. While conventional thin-film-composite membranes cannot be backwashed, the membrane used here (prepared by dynamic layer-by-layer (LbL) coating of a cationic polymer – poly diallyl dimethylammonium chloride and anionic polymer - polystyrene sulfonate (six bi-layer) on a polyethersulfone (PES) ultrafiltration (UF) hollow fibre) are mechanically robust. We filtered real tertiary effluent in lab-scale semi-batch mode with/without the automatic backwash protocol. Applying backwash with the NF permeate, we reached ~98% water recovery (concentration factor (CF) = 57), reducing the retentate volume to only 2% while controlling membrane fouling. Without backwashing, we reached only 94% (CF = 18) water recovery, representing a ~300% increase in the retentate volume. Furthermore, we obtained better permeate water quality for filtrations with backwashing, highlighting its effectiveness in high recovery effluent NF. Our membrane chemical cleaning protocol, applied after each high-recovery experiment, effectively restored water permeability and salt rejection (~ 90-98%) with/without backwash. We demonstrated that backwashing HF-PEMM NF membrane effectively achieves high water recovery (≥95%) while significantly minimizing retentate waste volume.

3:00 pm (Yellow Hammer, Wednesday)

Oral 159 - A membrane-based multi-barrier approach to potable water reuse and waste heat recovery

Andrea Achilli (University of Arizona)*; Zachary Binger (The University of Arizona); Mukta Hardikar (The University of Arizona); Sage Hiibel (University of Nevada, Reno); Eric Marchand (University of Nevada, Reno); Amy Childress (University of Southern California)

Onsite wastewater treatment and potable reuse technologies represent significant electrical energy consumption and operational costs at Department of Defense (DoD) fixed installations. In this work, a membrane bioreactor – membrane distillation (MBR-MD) system is proposed to accomplish high quality water reuse and represent an opportunity for energy independence and water supply security. This MBR system aims to accomplish treatment of wastewater to potable reuse standards while maintaining low electrical energy demand.

In MBR applications, external pressure (or vacuum) induces the flow of water across a porous membrane where suspended solids and other macromolecules are rejected by the membrane. In the proposed MBR system, wastewater is fed into a bioreactor and an ultrafiltration (UF) membrane provides high rejection of contaminants in the wastewater. The MBR filtrate is then processed by the membrane distillation (MD) stage that generates high-quality product water. MD involves the transport of mass and heat through a hydrophobic, microporous membrane. In MD, a feed solution at elevated temperature is in contact with one side of the membrane and the permeate is collected on the opposite side of the membrane; the temperature difference across the membrane induces a vapor pressure gradient for mass transfer. Compared to conventional distillation methods, MD requires only small temperature differences achievable using low-grade or waste heat sources.

This treatment scheme represents a multi-barrier system that includes biological removal, membrane separation, and thermal destruction. This is particularly important when microbial and trace organic contaminants (pharmaceuticals or personal care products) are present. Using waste heat existing on-site at DoD installations, this MBR-MD system can produce potable-quality water with minimal electrical energy input and minimal waste, resulting in lower operational costs as well as water and energy savings.

3:30 pm (Yellow Hammer, Wednesday)
Oral 160 - Transport mechanisms of fouled engineering-scale water reuse reverse osmosis membranes

Bianca M. Souza Chaves (The University of Arizona)*; Mohammed Alhussaini (The university of Arizona); Varinia V Felix (The University of Arizona); Andrea Achilli (University of Arizona)

The solution-diffusion (SD) model is widely used to describe water and salt transport in reverse osmosis (RO), however it may not describe all the complex transport mechanisms of water and solute molecules through the membrane. Pores or defects on the membrane surface can severely compromise selectivity. An SD-with-pore-flow model could best describe water flux and solute rejection. An engineering-scale RO system treating fifteen thousand gallons per day of reclaimed water operated continuously for two years before two fouled RO membrane modules were removed for analysis. The TW30-4040 RO spiral-wound elements were unrolled and cut for controlled chlorination in a bench-scale system to simulate membrane defects. The goal was to elucidate the oxidized membrane degradation pathway and the mechanism of salt/organic permeation by exposing the end-of-life RO membranes to different chlorine exposure doses (ED), from 1,000 to 48,000 ppm-h. Water, conductivity, ion, and organic transport were related to the ED. Membrane autopsy was investigated by contact angle, zeta potential, FTIR, and XPS analysis. The water permeability increased as the chlorine dose increased, with ~5-fold increase after 6,000 ppm-h (~2.5 m/s psi) compared to the non-chlorinated membrane, indicating that the membrane surface was modified. The polyamide layer was completely removed after 48,000 ppm-h (same initial and final conductivity). Solute permeability was calculated for ions and organics at different chlorine doses and the pore-flow permeability coefficient (Ld) was ~9-fold higher for monovalent ions compared to organics. A linear correlation in terms of log reduction values (LRV) was observed for sulfate and total fluorescence, calcium and sodium, and sulfate and sodium after 2,000 ppm-h. Results suggest that SD-with-pore-flow model could best describe the solute flux of fouled RO membranes and data will be discussed in the presentation.

4:00 pm (Yellow Hammer, Wednesday)
Oral 161 - Selective removal of selenium from saline water using different MOFs and MOF-incorporated NF membrane

Sweta A Modak (The University of Alabama)*; Milad Rabbani Esfahani (The University of Alabama)

Selenium shows the tendency of bioaccumulation in the food chain and shortens the life span of aquatic and wild animals. Concentration of selenium greater than 0.05 ppm is hazardous for human health, and hence, it is crucial to remove it from drinking water. In addition, selenium is one of the components of the produced water (the waste stream in the oil extraction process), which limits the reuse of the treated stream. Metal organic frameworks (MOFs) are gaining significant attention recently as adsorbents due to their high porosity, adsorptivity, and selectivity. In this study, the nickel-based ionic MOF (Ni-iMOF), zirconium-based MOF (MOF-808), iron MOF(Fe-MOF), copper MOF (Cu-MOF) and graphene oxide immobilized MOFs (GO-Fe-MOF and GO-Cu-MOF) were synthesized via solvothermal method using different metal salts and ligands and examined for the removal of selenium from water and saline water. Ni-iMOF, MOF-808 and Fe-MOF could remove selenite SeO32- (upto 75 mg/g) and selenate SeO42- (upto 43 mg/g) from water based on ion exchange and coordination adsorption mechanisms in the presence of competing ions, including SO32- and SO42-. Based on the acceptable removal performance of MOF-808 and Fe MOF, they were selected for immobilizing on the TFC membrane for the generation of MOF-adsorptive membranes. The mentioned MOFs were immobilized on the polyamide layer through chemical grafting. The Fe-MOF membrane showed 6.63 LMH/bar permeability with enhanced selenium removal (about 88%). The effect of MOFs on the physicochemical properties of MOF-NF membrane and the selective performance along with the salt rejection of the fabricated MOF-NF membrane at different operation conditions (temperature (25-45°C), pH (2-12) and varying ionic strength (SO32- and SO42-)) are studied.

4:30 pm (Yellow Hammer, Wednesday)
Oral 162 - None

Oral 28 – Bioinspired and Biomimetic Membranes

2:00 pm (Black Warrior, Wednesday)
Oral 163 - Tunable membranes incorporating artificial water channels for high-performance water

reverse osmosis desalination

Mihail Barboiu (Institut Europeen des Membranes)*

Membrane-based technologies have a tremendous role in water purification and desalination. Inspired by biological proteins, artificial water channels (AWC) have been proposed to overcome the permeability/selectivity trade-off of desalination processes. Promising strategies exploiting the AWC with Å-scale selectivity, have revealed their impressive performances when embedded in bilayer membranes. Herein, we demonstrate that self-assembled imidazole-quartet (I-quartet) AWCs are macroscopically incorporated within industrially relevant reverse osmosis membranes. In particular, we explore the best combination between I-quartet AWC and m-phenylenediamine (MPD) monomer to achieve a seamless incorporation of AWC in a defect-free polyamide membrane. The performance of the membranes is evaluated by cross-flow filtration under real reverse osmosis conditions (65 bar or 15-20 bar of applied pressure) by filtration of brackish feed streams. The optimized bio-inspired membranes achieve an unprecedented improvement, resulting in more than twice (up to 2.5 L m−2h−1 bar-1 or 6.9 L m−2h−1 bar-1) water permeance of analogous commercial membranes, while maintaining excellent NaCl rejection (>99.5%). They show also excellent performance in the purification of low-salinity water under low-pressure conditions (6 bar of applied pressure) with fluxes up to 35 L m−2 h−1 and 97.5-99.3% observed rejection.

1. M. Di Vincenzo, A. Tiraferri, V.-E. Musteata, S.Chisca, R.Sougrat, L.-B. Huang, S. P. Nunes, M. Barboiu,* Nature Nanotechnol. 2021, 16, 190-196
2. L.-B. Huang, M. Di Vincenzo, M. Göktuğ Ahunbay, A. van der Lee, D. Cot, S. Cerneaux, G. Maurin*, M. Barboiu*, J. Am. Chem. Soc. 2021, 143, 14386-14393.
3. M. Di Vincenzo, A. Tiraferri, V.-E. Musteata, S. Chisca, M. Deleanu, F. Ricceri, D. Cot, S. P. Nunes, M. Barboiu*, Proc. Natl. Acad. Sci., 2021, 118(37), e2022200118.

2:30 pm (Black Warrior, Wednesday)

Oral 164 - Development of Nanofiltration membranes incorporating Artificial Water Channel

Omar Samhari (Université de Montpellier)*; Mihail BARBOIU (Institut Europeen des Membranes)

Membrane-based technologies have a wide role in water purification and desalination. Artificial Water Chanel (AWCs) are synthetic aquaporins that can offer high water permeation with high selectivity. This work demonstrates that self-assembled imidazolequartet (I-quartet) can be incorporated into polyamide Nanofiltration membrane in order to increase the performances of the membranes. Polyamide Nanofiltration membranes were fabricated via interfacial polymerization with AWCs and tested in desalination for Brackishwater. An optimization was done to explore the best combination between AWCs and Piperazine monomer to achieve a maximum of performances in a defect-free polyamide membrane. The performance of the elaborated membranes was evaluated in cross-flow filtration, Nanofiltration condition (12-10 bar) of brackish feed streams. As expected, the optimized AWCs membrane achieved an insane improvement, the permeability dramatically increases in the bioinspired membranes resulting in more than twice the permeability of the reference membranes (44 L·m−2·h−1·bar−1) while the NaCl rejection was maintained above 62%.

3:00 pm (Black Warrior, Wednesday)

Oral 165 - Effect of Nanopatterning on E. Coli Fouling and Concentration Polarization

Lauren M Ward (The University of Alabama)*; Georgie Fickling (The University of Alabama); Urmi Roy (The University of Alabama); Steven T Weinman (The University of Alabama); Rushabh Shah (University of Massachusetts Amherst); Jessica Schiffman (University of Massachusetts Amherst)

Surface patterning is a recent growing topic in the membrane community. Membrane patterning became an interest after success was had in biomedical applications with the Sharklet® pattern that mimics the form and function of shark skin. These surface patterns disrupt the fluid flow at the membrane surface, creating localized turbulence or eddies that reduce fouling and make the membranes easier to clean. Fouling is when there is a blockage of the membrane pores or build-up of particles on the membrane surface, which decreases rejection and water transport through the membrane. Concentration polarization occurs when there is there is a build-up of salt ions right above the membrane surface due to the salt ions being rejected while water permeates through the membrane. By adding surface patterns, it is hypothesized that (1) concentration polarization will be disrupted and (2) bacteria fouling will be decreased on the membrane surface. Therefore, if methods can be developed that reduce or prevent concentration polarization and bacteria fouling, membranes will be more economical to operate. In this study, a line-and-groove nanopattern was applied to a nanofiltration (NF) membrane and an ultrafiltration (UF) membrane with a total patterned area of 6.4 cm2 to test the effects on concentration polarization and bacteria fouling respectively. The pattern increases the surface area of the membrane, while having a statistically significant increase of the salt flux of the membrane. NF membranes were tested in a direct-flow cell with deionized water, 2,000 ppm Na_2SO_4, and 10,000 ppm Na_2SO_4 solutions. The UF membranes were tested in a dead-end cell with water and E. Coli solutions to test for fouling and cleaning capabilities. The results show that the nanopatterns worsened concentration polarization but improved the cleanability of the membranes.

3:30 pm (Black Warrior, Wednesday)

Oral 166 - Biomimetic Supramolecular Channels for Sustainable Extraction of Rare Earth Elements

Harekrushna Behera (The University of Texas at Austin)*; Manish Kumar (University of Texas at Austin); Hyeonji Oh (University of Texas at Austin); Laxmicharan Samineni (University of Texas at Austin); Tyler J Duncan (The University of Texas at Austin); Venkat Ganesan (The University of Texas at Austin)

Rare earth elements (REEs) are crucial for modern industry and high-tech devices, including renewable energy and energy efficient infrastructure materials including in batteries for electric cars, energy-efficient lighting, display panels, and magnets for wind turbines. The US Department of Energy has identified europium as one of the five critical REEs for low-carbon and green energy technologies, which are vulnerable to supply chain disruptions. However, the domestic production of these critical elements has significantly declined, and their extraction and purification from natural ores are costly, energy-intensive, and damaging to the environment. Efficient separation technologies for the recovery of REEs from electronic wastes are currently lacking, which makes the recycling process challenging. Size cannot be a significant discriminating factor for the separation and purification of lanthanides (Ln(III)) elements, as they have nearly identical sizes. To address these challenges, we proposed an affinity-based biomimetic separation technique that utilizes specific ligands for Ln(III). Here, we developed supramolecular channels based on a pillar[5]arene scaffold with appended diphenyl phosphine oxide ligands (DPP) that demonstrate remarkable selectivity for Ln(III) over common monovalent ions (≥18) during electrochemically driven (voltage clamp) transport measurements. These channels exhibited promising selectivity towards medium Ln(III), such as Europium ions (20 times) over heavy Ln(III) (Yb^{3+}) and 41 times over light Ln(III) (La^{3+}), while excluding all measured mono- and divalent ions, including protons. The high selectivity of the channels is attributed to their DPP functional groups, which can dehydrate medium Ln(III), while the oxygen atom located just above the central ring makes a selectivity filter and decide their selectivity. Overall, this class of channels could serve as excellent building blocks for sustainable lanthanide separation membranes.

4:00 pm (Black Warrior, Wednesday)

Oral 167 - Biomimetic membranes with high water vapor permeance and protective capacity for next generation breathable protective fabrics

Hyeonji Oh (University of Texas at Austin)*; Yu-Ming Tu (MIT); Behzad Mehrafrooz (UIUC); Veda Sheersh Boorla (Pennsylvania State University); Harekrushna Behera (University of Texas at Austin); Costas Maranas (Pennsylvania State University); Aleksei Aksimentiev (University of Illinois at Urbana-Champaign); Manish Kumar (University of Texas at Austin)

An ability to transport water vapor at high rate is crucial in many applications, including breathable protective fabrics. Existing protective fabrics are lack of this ability despite providing effective protection against harmful chemicals that are used in our daily lives, warfare, and terrorism. An enhanced capacity to transport water (sweat) vapor helps natural body temperature regulation and increase active work hour for first responders by reducing heat stress. Here, we developed biomimetic membranes that have both high water vapor permeance and protective capacity, utilizing 2D nanosheets of beta-barrel biological channel, outer membrane protein F (OmpF). The membranes show excellent water vapor permeance of 602 ± 92 g h^{-1} m^{-2} that is 6.2 and 1.4 times higher compared to the current state-of-the-art water vapor permeable carbon nanotube membranes and Pertex shield, a commercial breathable fabric. High protective capacity against various range of small molecules including dye (304 to 972 Da) and model biological agents (MS2 bacteriophage, 2MDa) was also demonstrated. Interestingly, high water resistance in liquid phase was acquired during dehydration process. These unique properties of the membrane suggest promising applications in a variety of areas beyond breathable protective fabrics including barrier for water harvesting materials, medical devices, and distillation/pervaporation.

4:30 pm (Black Warrior, Wednesday)

Oral 168 – None

Oral 29 – Membrane and Module Optimization

2:00 pm (Capstone, Wednesday)

Oral 169 - Random polyampholyte amphiphilic copolymers for membranes with tunable selectivity and excellent fouling resistance

Ayse Asatekin (Tufts)*; Luca Mazzaferro (Tufts University)

Two of the biggest challenges in membrane science have been the prevention of fouling and the precision control of membrane selectivity. Designing novel polymeric materials with controlled surface chemistry and nanosctructure has the potential to overcome the limitations of existing membrane chemistries that curtail efforts to address these two problems. New membrane materials should ideally be capable of addressing various separation challenges and applications, allowing precise control over effective pore size, electrostatic interactions, and pore chemistry along with fouling resistance. Here, we describe a family of self-assembling copolymers that exhibit these key features when used as the selective layers of thin film composite membranes, random polyampholyte amphiphilic copolymers (r-PACs). r-PACs are statistical terpolymers comprising hydrophobic, anionic, and cationic groups. These copolymers self-assemble to create a bicontinuous network of ionic nanodomains that serve as effective nanochannels for water permeation, while the hydrophobic nanodomains preserve the structural integrity of the system. This copolymer design approach enables precise selectivity control by changing the chemistry and ratios of the three monomers. For example, membranes with sodium sulfate rejections varying from 5% to 93% can be manufactured with no significant change in pore size or fouling resistance. The ampholytic nature of the material leads to excellent fouling resistance, particularly when the anionic and cationic repeat units are at a 1:1 molar ratio. Membranes developed here have potential applications in wastewater treatment, ion separations, chemical and biological separations.

2:30 pm (Capstone, Wednesday)

Oral 170 - Co-transport of methanol and formate in crosslinked phenyl acrylate-based ion exchange membranes: Effect of steric hindrance

Jung Min Kim (University of Virginia)*; Yihung Lin (Auburn University); Sean M Bannon (University of Virginia); Geoffrey M Geise (University of Virginia); Bryan S Beckingham (Auburn University)

Understanding the co-transport behavior of CO2 reduction products (methanol and formate) in ion exchange membranes (IEMs) is of interest for CO2 reduction cells (CO2RCs). The role of an IEM in a typical CO2RC is to suppress the crossover of all CO2 reduction products while allowing the transport of electrolytes. Tuning the polymer rigidity of the membrane is a key contributor to such highly controlled transport of organic solutes in a dense hydrated membrane. Here, we investigate the co-transport behavior of methanol and formate in a series of tough phenyl acrylate-based crosslinked IEMs. We then investigate the effects of a structural modification on co-transport behavior by introducing quaternary carbons within the membrane. First, we measured diffusive permeabilities and sorption coefficients for methanol, formate, and a combination thereof. We then measured the relative permittivity properties, which are related to selective transport (of electrolytes over CO2 reduction products) through the membrane, of the swollen films to determine if the presence of the organic solutes impacted the nature of the hydrogen bond network within the hydrated IEMs. We observed that the films with methyl group incorporation have effectively constant relative permittivity values when exposed to solutions containing methanol, formate, and a combination thereof. These findings may assist in designing membranes for applications, including CO2 reduction cells and water-organic separations.

3:00 pm (Capstone, Wednesday)

Oral 171 - Acoustic actuated 3D printed bubble-trapping structures for in-situ mitigation of membrane fouling

Kieran R Fung (Membrane Science, Engineering, and Technology (MAST) Center, Department of Mechanical Engineering, University of Colorado Boulder)*; Shouhong Fan (University of Colorado at Boulder); Jason P. Killgore (Applied Chemicals and Materials Division, National Institute of Standards and Technology, Boulder, CO); Yifu Ding (Membrane Science, Engineering, and Technology (MAST) Center, Department of Mechanical Engineering, University of Colorado Boulder); Xiaoyun Ding (Department of Mechanical Engineering, University of Colorado Boulder)

Acoustic streaming bubbles have garnered significant attention over the last decade for its ability to instantaneously generate turbulence and enhance mass transport. Upon acoustic actuation, mechanical perturbations oscillate the gas-liquid-interface and generate steady streaming effects. Prior studies have the demonstrated the utility of acoustic streaming bubbles in applications like pumping, mixing, and propulsion. In this work, we present a three-dimensionally printed elastomeric structure for acoustic streaming in bulk fluid. The structure exhibits a spherical cavity at the tip, which is used for trapping bubbles when submerged in aqueous environments. It is shown that the streaming velocity is proportional to the amplitude of oscillation, which is maximized at the resonant frequency of the bubble (~120 mm/s). The generated vorticial flow is also seen to persist millimeters past the structure tip, suggesting strong promise for laminar flow disruptions. Accordingly, the proposed bubble-trapping structures are integrated into a silica particle liquid separation process to mitigate membrane fouling and cake layer buildup. Arrays of the structures are fixed onto the feed plate of the membrane flow cell that point towards the membrane rejection surface to perform acoustic cleaning. In this presentation, we explore the effectiveness of bubble-trapping structures to act as a novel mechanism to mitigate membrane fouling.

3:30 pm (Capstone, Wednesday)
Oral 172 - High-performance large-area polymeric membranes with chlorine tolerance for water purification by pressure-driven distillation

Duong T. Nguyen (University of Colorado Boulder)*; Kian P Lopez (University of Colorado Boulder); Shouhong Fan (University of Colorado Boulder); Yifu Ding (Membrane Science, Engineering, and Technology (MAST) Center, Department of Mechanical Engineering, University of Colorado Boulder); Jongho Lee (University of British Columbia); Anthony Straub (University of Colorado Boulder)

State-of-the-art reverse osmosis membranes poorly remove low-molecular-weight, neutral solutes and are prone to degradation by chemical oxidants. Pressure-driven distillation, a novel membrane process using applied hydraulic pressure to drive water vapor transport, enables water separation with ultrahigh selectivity and oxidation tolerance and potentially reduces cost and energy consumption for desalination and water reuse. The process has been demonstrated in preliminary studies using costly and delicate ceramic membranes. However, these materials pose challenges for scaling up due to limited surface area, high cost, and deformation in pressurized modules. Here, we present a scalable approach to fabricate high-performance robust membranes by controlled metal sputtering and hydrophobic modification on different hydrophilic polymeric substrates. The thickness of the selective metal layer can be tuned by adjusting sputtering angle and time, which allows for increased water throughput without sacrificing solute selectivity. The membranes were fabricated on a polyvinylidene fluoride substrate and can operate under a high hydraulic pressure of 34.5 bar, offering a water permeability of 3.3 L m-2h-1bar-1 and a sodium chloride rejection of more than 98.5%. In addition, the membranes exhibited near-complete removal of hazardous small neutral contaminants including boron, N-nitrosodimethylamine and haloacetonitrile, and unchanged desalination performance after severe exposure to high doses of chlorine. Such advantages of the membranes will facilitate the implementation of pressure-driven distillation and pave the way for low-cost and energy-efficient water purification.

4:00 pm (Capstone, Wednesday)
Oral 173 - Graphene Oxide and Sulfonated graphene oxide embedded Polyamide-polysulfone based Thin-film Nanocomposite Sea water Reverse Osmosis membranes

Asim K Ghosh (Bhabha Atomic research centre, Trombay, Mumbai)*; Rutuja Bhoje (Institute of Chemical Technology); Parag Nemade (Institute of chemical Technology)

The productivity of a membrane in terms of higher flux and salt rejection are two important parameters for desired reverse osmosis (RO) membranes. In addition, chlorine and biofouling resistance properties to the RO membrane provide added advantages particularly to enhance the membrane's life. The thin-film nanocomposite (TFN) based seawater RO membranes commercially available (SW 400 ES & SW 400 SR G2 by LG M/s. LG Chem) have the highest water flow and salt rejection among any RO membrane on the market but are neither chlorine tolerant nor biofouling resistant. To fabricate a high-performance TFN-RO membrane, graphene oxide (GO) and sulfonated graphene oxide (SGO) nanoparticles were synthesized and incorporated into the active polyamide layer of the composite membranes in this study. Composite polyamide membranes were synthesized over polysulfone support membranes by in-situ interfacial polymerization using MPD in an aqueous solution and TMC in the Isopar-G solvent. GO and SGO are taken in an aqueous amine solution with concentrations ranging from 0.0025 to 0.01%. Due to the intrinsic features of GO and SGO nanoparticles, their incorporation into pristine polyamide thin-film composite (TFC) membranes dramatically improves the performance, chlorine tolerance, and biofouling resistance properties. Both TFC and GO/SGO-modified TFN membranes were characterized in terms of zeta potential, hydrophilicity, FE-SEM, surface average roughness, and FTIR analyses. The performance of TFC & TFN membranes was evaluated under seawater reverse osmosis (SWRO) testing conditions. The flux increase in TFN membranes is around 42% and 48% for GO and SGO-modified membranes respectively compared to TFC membranes with a marginal decrease in salt rejection. The membranes were subjected to 2*10-7 E. coli bacterial solution for 3 hr and 300 ppm NaOCl solution for 24hrs and demonstrated the increased anti-biofouling and chorine resistance capabilities of TFN membranes over pristine TFC.

4:30 pm (Capstone, Wednesday)

Oral 174 - Sweeping Gas membrane distillation for rapid heat dissipation

Waqas Alam (Purdue)*; David M Warsinger (Purdue); Hamid Juybari (Purdue University)

Membrane distillation (MD) is a potentially suitable process for treating reverse osmosis concentrate to produce distillate and reduce volume of concentrate to be disposed. High rejection of contaminants is an essential parameter for potable reuse of water treated in MD. Theoretically, MD can achieve 100% rejection of non-volatile components such as solutes and pathogens because only the vapor phase permeates through the membrane. Rejection of organics and salts has been reported in a wide range (90-99.99%) but its resolution is limited by the solubility of the contaminant in the feed and the detection limit of contaminants in the distillate. Also, thus far, pathogen rejection in pilot-scale MD has not been investigated. At the pilot-scale, the larger membrane area increases probability of transport of contaminants through pore defects while it decreases the water flux and dilution of contaminants in the distillate. Studying rejection of pathogens also overcomes the solubility limit of the contaminant in the feed. In this work, membrane rejection of non-volatile components in a pilot-scale air-gap MD (AGMD) element is quantified using bacteriophages. Reclaimed wastewater was used as feed and was spiked with two bacteriophages MS2 and PhiX174 at approximately $4x10^5$ PFU/mL concentration. Membrane rejection of bacteriophages, cations and organics was quantified at a constant flowrate and temperature difference between the feed and distillate (40/20 °C) and low and high vacuum in the air gap. Membrane rejection of bacteriophages was observed to be 0.5-log10 higher than cation rejection. Rejection of organics, inorganics and bacteriophages reduced with increase in vacuum in the air gap, which suggests presence of pore flow through pore defects. Interestingly, exposure to high feed temperature (70 °C) did not lead to an instantaneous decrease in feed virus concentration suggesting that longer operating times are required to decrease virus concentration in the feed.

NAMS 2023 - SPONSORS

DIAMOND LEVEL

National Science Foundation

The University of Alabama College of Engineering

Department of Chemical and Biological Engineering

GOLD LEVEL

Anton Paar

Membrane Science, Engineering & Technology (MAST) Center

MilliporeSigma

NL Chemical Technology

Prometheus Fuels

SILVER LEVEL

Chevron

Dow

Membrane Technology & Research

Osmoses

Pall/Cytiva

Honeywell - Universal Oil Products

ZwitterCo

BRONZE LEVEL

Air Liquide

Air Products

Ameridia Innovative Solutions Inc.

Bayer Crop Science

Evonik Chemicals

Georgia Institute of Technology

MDPI AG (Membranes Journals)

Sterlitech

Veolia

Gale & DB Battacharyya

Alabama Water Institute

EXHIBITORS

DataPhysics Instruments

Hiden Isochema

Porometer

FRIEND OF NAMS

Keith Murphy

NSF

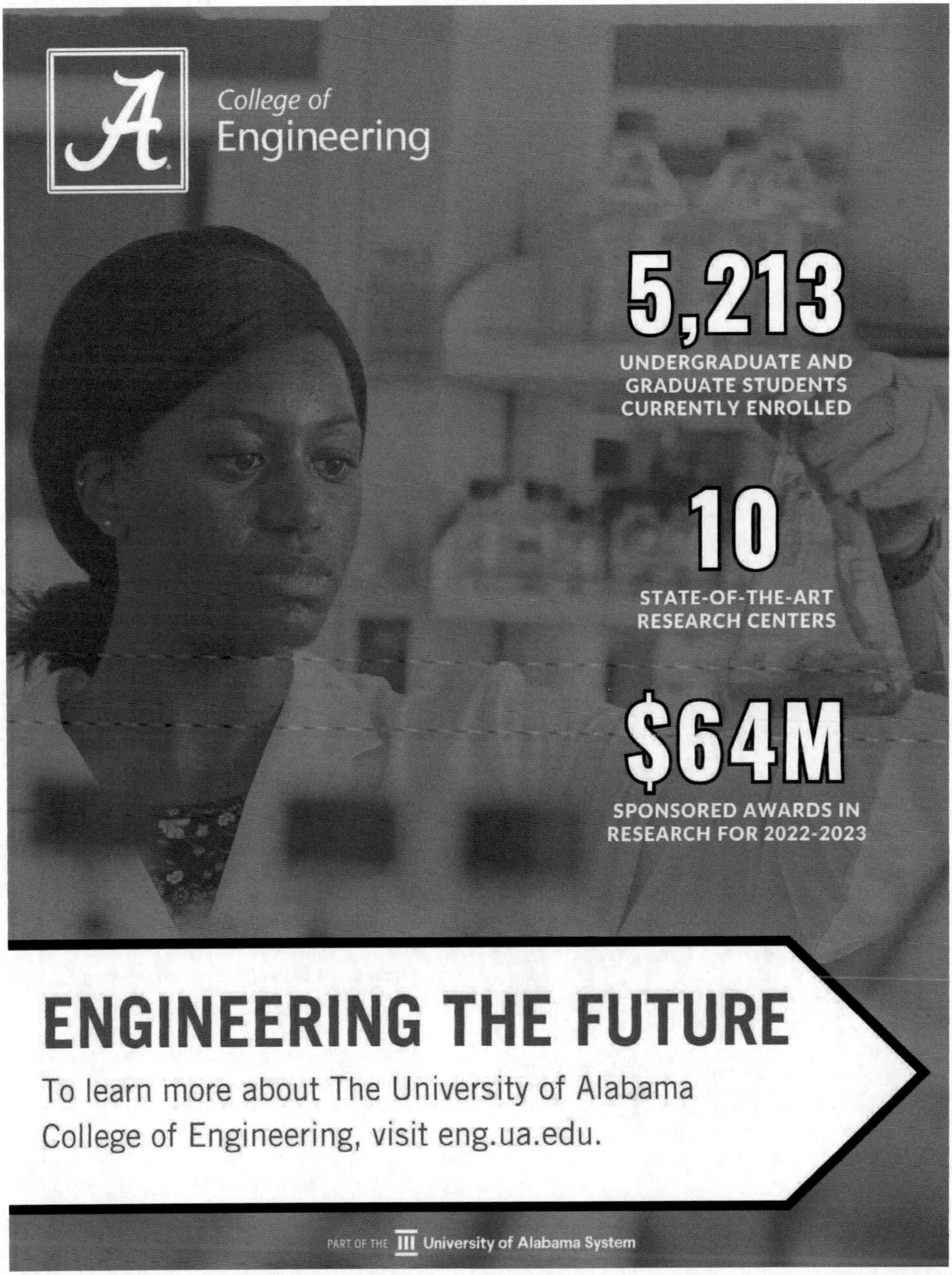
College of
Engineering
5,213
UNDERGRADUATE AND
GRADUATE STUDENTS
CURRENTLY ENROLLED
10
STATE-OF-THE-ART
RESEARCH CENTERS
$64M
SPONSORED AWARDS IN
RESEARCH FOR 2022-2023
ENGINEERING THE FUTURE
To learn more about The University of Alabama
College of Engineering, visit eng.ua.edu.
PART OF THE University of Alabama System

SurPASS 3

SURFACE ZETA POTENTIAL

Anton Paar

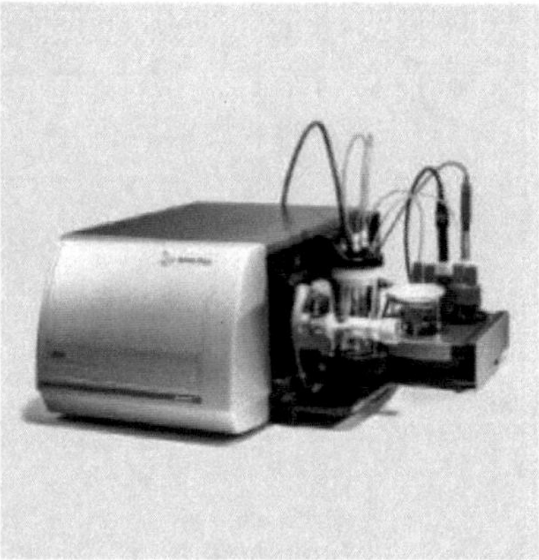

- Measure the zeta potential of membranes, powders, coatings and more
- Analyze a wide range of sample geometries and sizes
- Integrated titration unit

Learn More

NSF I/UCRC • Membrane Science, Engineering & Technology Center

MAST

A global leader in membrane research for over 30 years.

The Membrane Science, Engineering and Technology (MAST) Center has been conducting industrially relevant precompetitive research for over 30 years. As a collaborative effort between leading membrane industry sponsors and University researchers, the center focuses on developing innovative solutions to separations challenges. As an NSF and University Cooperative Research Center, MAST consists of four partner institutions: The University of Arkansas, The University of Colorado Boulder, The New Jersey Institute of Technology, and Pennsylvania State University. To learn more visit us at mastcenter.org.

NAMS 2023 – SPONSORS GOLD LEVEL

NL Chemical Technology Inc.

www.nlchem.com

- Broad Membrane Product Portfolio (RO/NF)
- Proprietary Technology Licensing
- Turnkey Transfer of Membrane Manufacturing Technology
- Six-sigma Quality Management

Developing membranes

from laboratory to commercial products

For business and employment opportunities, please contact:
Dr. Robin Wang e-mail: robin.wang@nlchem.com

NAMS 2023 – SPONSORS SILVER LEVEL

DOW®

Membrane Solutions
for Industrial Separations

MTR
MEMBRANE
TECHNOLOGY & RESEARCH

OSMOSES

Air Liquide

AIR PRODUCTS Membrane Solutions

Gale & DB Bhattacharyya

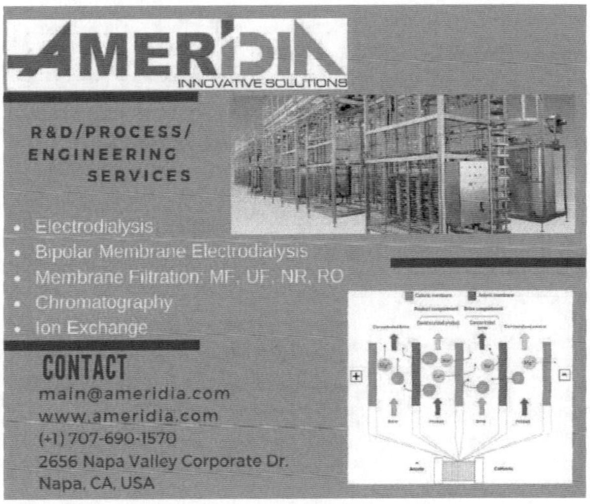

NAMS 2023 – SPONSORS BRONZE LEVEL

BAYER

EVONIK
Leading Beyond Chemistry

Georgia Tech | ISSEC

The **Integrated Separations Science and Engineering Center at Georgia Tech** (ISSEC) is designed to foster collaborative & interdisciplinary solutions to separation challenges by promoting research partnerships not only among faculty from many units but also partnerships with industry. These collaborations results in new separation methods that are sustainable and frequently implemented by our industrial partners. Equally important, ISSEC helps support students and works with industry and government to learn the fundamentals of separations science and to apply them to real world applications.

membranes

IMPACT FACTOR 4.562 | Indexed in PubMed

An Open Access Journal by MDPI

Membranes (ISSN 2077-0375) is an international, peer-reviewed, open access journal. The journal covers broad science- and technology-related aspects of both biological and non-biological membranes, published monthly online by MDPI.

Author Benefits

4.562
Impact Factor

Q1
Journal Rank: Q1 (*Polymer Science*)

12,1
days
Submission to First Decision

Open access
Unlimited and free access for readers

MDPI Academic Open Access Publishing since 1996

Membranes Editorial Office

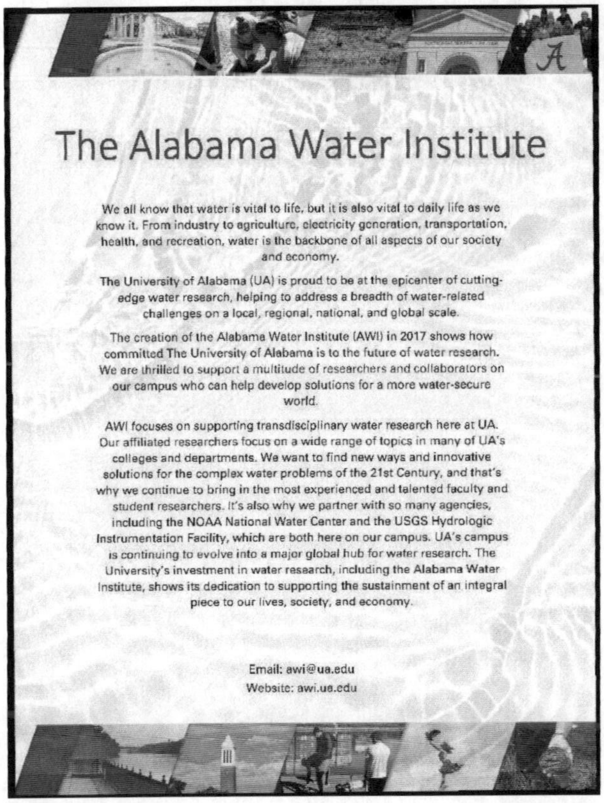

Thanks to All Our Sponsors for Their Continued and Generous Support.

NAMS 2023 - SPONSORS

DIAMOND LEVEL

NSF

College of Engineering
Chemical and Biological Engineering

GOLD LEVEL

Anton Paar

MilliporeSigma

NSF I/UCRC • Membrane Science, Engineering & Technology Center
MAST

SILVER LEVEL

Chevron

DOW

Honeywell UOP

Pall+Cytiva
Turns out, we have chemistry
Perfect match
cytiva

MTR
MEMBRANE
TECHNOLOGY & RESEARCH

OSMOSES

ZWITTERCO

BRONZE LEVEL

Air Liquide

STERLITECH
Corporation

Gale & DB Bhattacharyya

AMERIDIA
INNOVATIVE SOLUTIONS

BAYER

membranes

EVONIK
Leading Beyond Chemistry

Georgia Tech ISSEC

THE UNIVERSITY OF ALABAMA | Alabama Water Institute

AIR PRODUCTS

VEOLIA